D0945033

THE METHUEN EEC SERIES
General Editor: R. A. Butlin

Energy in Europe

for Christopher and Lawrence

Other books in this series:

Competition and Industrial Policy in the European Community
Dennis Swann

The Common Fisheries Policy of the European Community
Mark Wise

The Common Agricultural Policy – Past, Present and Future
Brian E. Hill

A Rural Policy for the EEC?
Hugh Clout

Forthcoming:

Transport Policy in the EEC
J. Whitelegg

Energy in Europe
Issues and policies

THOMAS G. WEYMAN-JONES

METHUEN
LONDON AND NEW YORK

First published in 1986 by
Methuen & Co. Ltd
11 New Fetter Lane, London EC4P 4EE

Published in the USA by
Methuen & Co.
in association with Methuen, Inc.
29 West 35th Street, New York NY 10001

Typeset by Graphicraft Typesetters Ltd., Hong Kong
Printed in Great British by Richard Clay,
Bungay, Suffolk.

British Library Cataloguing in Publication Data

Weyman-Jones, Thomas G.
Energy in Europe issues and policies.—
(The Methuen EEC series)
1. Power resources—European Economic
Community countries
I. Title
333.79′094 HD9502.E862

ISBN 0-416-36590-6

Library of Congress Cataloging-in-Publication Data

Weyman-Jones, Thomas G., 1947–
Energy in Europe.

(The Methuen EEC series)
Bibliography: p.
Includes index.
1. Power resources—Europe. 2. Energy policy—
Europe. I. Title. II. Series.
TJ163.25.E85W48 1986 333.79′094 86–18169
ISBN 0-416-36590-6

Contents

Notes on acronyms and abbreviations used in the book

AFME	Agence Françaisê pour la Maîtrise de l'Energie
CAP	Common agricultural policy
cba	Cost–benefit analysis
CEGB	Central Electricity Generating Board
CEP	Common energy policy
CHP	Combined heat and power
c.i.f. price	The price of a commodity when *cost, insurance* and *freight* charges are included
COM documents	Data from EC documents
CPE	Centrally planned economies
EC Commission	The European Communities' Commission, the body of civil servants which serves all three Communities
EDF	Electricité de France
EEC	European Economic Community (1957). This is one of the three European Communities, the others being the European Coal and Steel Community, ECSC (1951) and the European Atomic Energy Community, Euratom (1957). 'The Community' is synonymous with the EEC.
Eurostat	The Statistical Office of the European Communities, part of the EC Commission
EUR9, EUR10	The collective member states of the EEC at different times in its development. EUR9: the EEC from 1971, i.e. Belgium, Denmark, West Germany, France, Ireland, Italy, Luxembourg, the Netherlands and the UK. EUR10: the EEC from 1981, EUR9 plus Greece. Eurostat frequently gives data for EUR10 for periods

	before 1981. EUR10 is taken as the basic collective for the EEC in this book
FBR	Fast breeder reactor
f.o.b. price	The price of a commodity when *free on board,* i.e. excluding shipping and insurance charges
GDP,GNP	GDP is gross domestic product, a measure of the total volume of final output produced in the economy. Three ways of calculating it may be used: the sum of all value added in economic activity, the sum of all domestic incomes earned less stock appreciation, or the sum of total final expenditures including net exports. GNP is gross national product, i.e. GDP plus net property income from abroad. National income is GNP less an allowance for capital consumption in economic activity
IASBB	Inflation adjusted structural budget balance
IEA	The International Energy Agency. This consists of the members of the OECD except for Finland, France and Iceland
IMF	International Monetary Fund
JET	Joint European Torus
LRMC	Long-run marginal cost
mbd	Million barrels per day
mtoe	Millions of tonnes of oil equivalent
NAC	Net avoidable cost
NEC	Net effective cost
nodc's	Non-oil developing countries
OECD	The Organization for Economic Cooperation and Development, consisting of the principal industrialized countries, including all of the EEC, Japan and the USA
OPEC	The Organization of Petroleum Exporting Countries
PWR	Pressurized water reactor
RUE	Rational use of energy
SEC documents	Data from EC staff working papers

Figures and tables

Figures

Tables

Preface

When Robin Butlin invited me to contribute to his series of books on the EEC, he was kind enough to suggest that a book on the absence of a Common Energy Policy (CEP) might be too short, and that I should consider the wider context of energy economics in a European dimension. This has turned out to be an enormously wide brief, and the present study only scratches the surface of what has been one of the most interesting and critical aspects of European economic and political development.

As with the other books in this series, it has been assumed that readers will come from a variety of backgrounds: European studies, politics, economics, resource analysis, geographical and business studies, along with private and public sector industries and administrations in Europe and elsewhere. Consequently, although the book is about economics, I have tried to minimize the use of economists' jargon. The discussion of the economic issues should be straightforward to any reader with a general interest in European developments. No specialist knowledge of energy is assumed, and indeed much of the book may be old hat to specialists in energy economics.

In European studies the phrase 'the Commission proposes and the Council disposes' is a cliché, but in the absence of anything but rather bland and infrequent disposals, I have taken the liberty of including in the discussion references to a considerable amount of the European Communities' (EC) Commission's work on monitoring member states' policies and making forecasts.

One topic is deliberately excluded: discussion of what a CEP could or should be, and no doubt readers familiar with other work on EEC energy will be grateful for this relief. On the other hand, because the book is subtitled 'issues and policies', considerable attention is paid to the way energy problems have affected the rest of the economy. In particular, I have included a chapter on the macroeconomic response to oil shocks because I believe energy has only been perceived as a source of crisis precisely because oil price rises reduced general living standards and contributed to higher unemployment. In general I have stuck to the facts of energy developments, but several issues of principle do arise: e.g. the nature of oil import taxes, the determination of energy demand, the procedures of nuclear power cost–benefit analysis and so on, and in such cases I have tried to set out some of the general conclusions reached by economics so that the reader has a clearer idea of why certain policies are proposed or may be contentious.

A particular point about sources needs to be made. Most of the official EEC data quoted in this book is produced by the Statistical Office of the European Communities, a source usually abbreviated to Eurostat. On occasion I have made use of other data from Commission documents (COM documents) or staff working papers (SEC documents) and, for convenience in tables and figures, used the term Eurostat also to apply to these data sources. The text makes clear the more precise source involved in each case.

My debts are considerable, especially to Robin Butlin, the series editor, for his patience and encouragement while the book was being written.

At Loughborough, anybody working in the EEC area has the inestimable benefit of Dennis Swann's pioneering work, and continuing contributions and advice, and I am very grateful to him for starting me off on this topic. My initial interest in EEC energy policy was also stimulated by some research funding by the Commission, which I am happy to acknowledge.

Mike Hopkins of the University Library, and an acknowledged expert on EEC policy documentation, helped me enormously with sources, and several of my colleagues sorted out my ideas on economics: Mike Fleming, Colm Kearney, David Llewellyn, Chris Milner and Eric Owen Smith. In particular I have had many discussions with Peter Barker whose knowledge of everything from cartel theory to uranium supply I have borrowed shamelessly. None of these is responsible for any errors or omissions which follow.

Su Spencer has once again done a superb and speedy job of turning my scrawl into type.

Most important of all, my wife Carol has continuously encouraged me in all my work, and to all of these, but especially the last, I am very grateful.

T. Weyman-Jones

1 Energy in Europe

An analytical framework

The two themes of this book: the European Economic Community (EEC) and energy, have each attracted considerable interest in their own right, and their conjunction has been the subject of much debate and policy analysis.

On the one hand, the EEC has come to be regarded as a powerful political and economic grouping on the world stage, containing as it does a large proportion of the world's democracies and with a combined gross national product (GNP) approximately equal to that of the USA. On the other, energy economics emerged in the 1970s as absolutely critical in determining developments in the international economy, and particularly in western Europe where more than 40 per cent of the energy consumed is imported. Together with the USA (whose primary energy consumption is nearly twice as large as that of the EEC), the European economies have been ranged as large-scale oil consumers against the market power of the oil producers' cartel formed by the Organization of Petroleum Exporting Countries (OPEC). Both these groups have, in turn, developed special political and economic interests with the third world, consisting largely of the non-oil developing countries (nodc's). The latter have found themselves squeezed between the pincers of high oil prices and a burgeoning debt crisis.

It is important, however, to develop a consistent framework in which to analyse the issues and policies connected with energy in Europe, and two underlying characteristics of such a framework need to be emphasized.

The first is that the initial perceptions of the so-called energy crisis of the early 1970s were largely misplaced. They coincided with a widespread realization that the fossil fuels which largely dominate world energy consumption are finite resources, although there is no credible economic evidence that resource exhaustion forms a significant economic and technological constraint in the foreseeable future. Were this not the case, economics suggests that real resource prices net of extraction cost would be rising steadily through time at a rate approximately equal to the real rate of return on private sector capital investment as consumers competed for depleting exhaustible resources. There appears to be no observable evidence that this is, or has been, the case. Consequently, it is essential to distinguish between the idea that many fuels are exhaustible, and the apparent fact that fuel markets are not at present showing any signs that large-scale exhaustion is near.

Nevertheless, this idea alone persuaded many people that resource scarcity and the need to devote special efforts to energy conservation should be the overwhelmingly dominant characteristics of economic policy. A decade later it is clear that the preoccupation with the idea that energy was a special type of resource (essential but about to disappear) obscured much of the sensible and necessary economic and policy analysis of the events in fuel markets in the 1970s. While energy remains a specialized concept to the physicist, it is much more sensible for economic analysis to concentrate on the day-to-day market behaviour of individual *fuels* – treating these as any other scarce commodity demanded by consumers and industrialists. What is true is that the market responsiveness of energy consumers is spread out over considerable periods of time because any particular fuel type has to be used in conjunction with a specific type of capital asset or manufacturing process, and is therefore conditioned by long-term decisions to invest in or scrap the associated capital equipment. It is clear nevertheless that the ordinary economics of supply and demand is all that is required for developing the consistent framework to study energy in Europe.

The second characteristic of such a framework is that there is no reason to believe that it has a natural or comfortable European dimension. Although energy in Europe is the focus of attention here, it is not obvious that all the countries of Europe, or of the EEC, face the same fuel market problems, or have any agreed objectives to follow in fuel markets. It is not even clear that they have any common interest in fuel market resource allocation at all, other than the fact that there is an economic community in Europe which has been 'shocked' by events in oil markets that have had large-scale macroeconomic effects. The EEC consists of both fuel consumers and fuel suppliers. Its members have developed, over the years, commitments to different fuel types (e.g.

nuclear electricity in France, nuclear energy and coal in Germany, coal and subsequently oil in the UK, natural gas in the Netherlands) and hence a desire to protect those industries and their employees from outside competition. Its members have a history of ties with different external fuel suppliers: e.g. France with Algeria, the UK with the Persian Gulf, and very different attitudes towards the link between foreign policy initiatives and fuel markets that emerged in the 1970s. France, for example, refused to join the Organization for Economic Cooperation and Development (OECD)-based International Energy Agency (IEA) although all the EEC countries were members, and the EEC has subsumed its oil stockpile policy into that of the IEA.

As a consequence of these facts, the joint theme, 'energy in Europe', is much better considered as the analysis of European fuel markets and the external shocks imposed on them. Policies are, and have been, naturally formed at the national level, and it will be surprising if there is ever any obvious Common Energy Policy (CEP) in the sense of a counterpart to the Common Agricultural Policy (CAP). Nevertheless both the IEA and more particularly the EC Commission have adopted policy statements and demand–supply projections that appear to have a common EEC basis, and it will be necessary to examine these. Such statements and projections can be thought of as energy economics commentaries, and in developing an analytical framework for energy in Europe, no profit is, at present, to be gained from looking further for a discernible CEP. The extent, and at the same time, the limitations of such a concept are clearly implicit in the following, slightly anguished comment from the 1985 Energy Commissioner, Mr Nicolas Mosar:[1]

> The point I wish to make is that there really is a Community energy policy, the key words of which are 'coordination of Member States' policies (energy objectives, energy pricing policy and crisis system for example), and reinforcement of national measures' whenever necessary.

Given that fossil fuel exhaustion and an obvious and well-documented CEP are two of the misconceptions that a study of energy in Europe can do without, how is the necessary analytical framework to be developed?

Since the theme of energy in Europe is only a statistical and institutional artefact arising from the fact that EEC countries have experienced fuel market disruptions, it is perhaps wisest to begin with a statistical framework. This is provided by looking at a summary energy balance table for the EEC (table 1.1, page 10) which indicates (a) the sources of primary energy (i.e. fuels as raw materials) available to the EEC and (b) the different market sectors in which different types of secondary energy (i.e. fuel supplies after refining or transformation, transmission and distribution) are consumed. This statistical framework indicates the overwhelming significance of imported oil in the EEC's fuel markets, and it is this *unavoidable dependence on imported oil* which has underlain most of the explicit policy actions by member states. In addition, the energy balance table highlights the role of the secondary energy supply industries for coal, gas, electricity and petroleum products. These industries are large scale and capital intensive but their commercial organization differs significantly across Europe. It includes the cartelized private sector suppliers of Germany, the integrated national energy undertakings in Italy, and the separate nationalized fuel supply industries in the UK. In one way or another, government control is a part of the industrial structure in these markets, and this continually raises a tension between the requirements of efficient fuel resource allocation, and the objectives of macroeconomic policy. Had the fuel markets been able to adjust to successive oil shocks without macroeconomic consequences, the energy crisis might not have been so celebrated, but the fuel market disruptions of the 1970s turned out to embody the largest macroeconomic shocks experienced by the European economies since the Second World War.

The macroeconomic impact of oil shocks is therefore the first substantive analytical issue that emerges from the statistical consideration of European energy markets. For convenience three oil shocks will be said to characterize the years 1973 to 1986:

First oil shock: 1973–4 OPEC real price rises for oil exports
Second oil shock: 1979–80 OPEC real price rises in the wake of the Iranian revolution
Third oil shock: 1982–6 erosion of the real oil price in world markets

However, it is doubtful whether the third is really a shock to the macroeconomic system rather than a delayed market response to the first two shocks. Since these were so large that fuel markets could not absorb them without large-scale impacts on the general price level, the rate of inflation and the levels of aggregate economic activity and employment, their final effect was partly determined by macro-economic policy measures. These can be categorized in two ways. Since the general level of prices is affected there is an *implicit* policy response because the real value of fiscal and monetary policy instruments that are fixed in nominal or money terms will have changed: examples are the existing levels of tax allowances and any specific taxes such as those on motor spirit, and the level of the money supply in an economy. There may also be an *explicit* policy response, perhaps to offset the implicit effect, or to reinforce it if it favourably affects some other policy option. An example of the latter is that the second oil shock in the form of a higher price level actually deflated the European economies and this deflation was in some cases reinforced by active fiscal and monetary policy deflation to reduce further the rate of inflation. This was in strong contrast to the response to the first oil shock, which largely tried to offset its recession impact.

Since the importance of energy economics emerged first in the form of oil price rises, a second important analytical issue in the framework is the *role of OPEC, and the attitudes of the EEC, individual member states, and other world energy market participants such as the IEA, USA and Japan towards OPEC's use of its cartel power.*

Three strands of opinion have emerged about the role of OPEC in world energy markets. The first is that OPEC – a collection of the principal oil exporters, numerically dominated by the Middle Eastern states, and dominated in terms of oil reserves by Saudi Arabia – can be regarded as just one more market cartel or collusive monopoly intent on the maximization of long-run profits, and, like most cartels, susceptible to internal economic conflicts which have possibly given it only a fairly short lease of life. Since the world oil market has exhibited many such examples of market power, this view of OPEC follows naturally from previous analyses. Although supported by many observers, it is fair to say that the leading proponent of this view has for a long time been Professor Morris Adelman of MIT. Adelman's argument is essentially based on the methodology of positive economics: the predictions of the monopolistic theory are consistent with OPEC's observed behaviour.

A second strand of opinion takes the position that even without OPEC, oil prices were set to rise because of the long-term expansion of energy demand observed in the 1960s and 1970s. Since this does not account solely for the observed fluctuations and the rapidity of price rises when they came, it is often combined with the view that OPEC itself, as an instrument rather than a cause of oil price rises, is not a cohesive cartel of like-minded profit maximizers, but consists of countries with differing objectives towards, and capacities for, rapid economic development. This has led to the development of quite complex models of OPEC behaviour although the success of their predictions is not clearly established.

A third view of OPEC, a development of the second, is that it is essentially a political rather than an economic grouping and, as such, foreign policy and diplomacy should be the key ingredients of energy policy. In particular this view focuses on the long-term political objectives of Saudi Arabia in order to explain its apparent willingness to accept reduced oil revenues in the early 1980s. The difficulty with this view is that it is designed to explain in retrospect all aspects of OPEC behaviour and hence is impossible to test by the usual procedure of confronting theoretical predictions with subsequent facts.

The oil shocks of the 1970s and 1980s turned out to have enormous impacts on the markets for other fuels, and on energy demand in general. Consequently, a third important analytical issue in the energy in Europe framework being set out here, is the *development of Europe's indigenous fuel production, its consumption of imports and its overall energy conservation.* Once the considerable lags in market response had been worked through, the oil price rises, and the sympathetic rises in other fuels which accompanied them, were found to have a much greater effect on energy consumption than many economists had initially predicted. The previous decades of relatively stable prices and largely uninterrupted economic growth made it difficult to infer from pre-1972 data just how responsive fuel consumption would be in the long run following substantial price rises. Economists had always argued, in the face of considerable opposition, that there would be a market response, but the consensus observed long-run price elasticity has turned out to be considerably larger than many had predicted. The second oil shock in particular has produced a very noticeable drop in world oil consumption confirming economic theory that the price elasticity itself rises the higher the price level from which any further price escalation occurs.

The four principal primary energy sources: oil, coal, nuclear electricity and natural gas, have all felt the effects of the oil shocks. Europe's own production of oil was growing with North Sea developments just at the start of the oil price rises, and much of the North Sea was commercialized by OPEC's actions. North Sea oil exported from the UK has largely gone into Europe and had a fundamental effect on the UK economy. Norway opted out of EEC membership in order to reinforce control over its own very substantial oil and gas reserves. Coal might have been expected to be the major beneficiary of the oil price rises at a time when its market share was steadily crumbling. In fact protection of indigenous coal supplies was an early preoccupation of the European Communities and also for the UK. However, economic incentives to increase coal use only highlighted the uncompetitiveness of much German, Belgian and UK deep-mined coal compared with the growing supplies on world markets of surface-mined coal from Australia and the USA, and the coal protection lobby has had a very rough ride, culminating in the UK coal strike of 1984–5 and the EEC Commission stance on removal of coal production subsidies.

In those member states that had emphatically switched from coal to oil in the 1950s and 1960s, particularly France, the oil shocks gave a huge impetus to the case for nuclear power. This in turn has engendered its own political opposition and a considerable debate on the economics of the cost-benefit analysis of very long-lived projects with uncertain spillover effects.

The oil shocks also coincided with the huge expansion in the market share of natural gas in Europe, again as a result of rapid depletion of North Sea fields. This in turn has been followed by rapid gas price rises to conserve supplies, until at the start of the 1980s the massive gas supplies which could be piped from the Urengoy fields of the USSR appeared with enormously attractive economic implications for both the EEC and the USSR but simultaneously large political uncertainties.

These three major analytical issues: macroeconomic policy response, relations with OPEC, and fuel market developments, all arising in Europe in the wake of the oil shocks have led to two major policy developments.

Firstly, a great deal of attention has had to be devoted to the economic basis of energy policy making, and in particular to the determination of energy prices. Amongst the competing arguments, the following can be noted. For efficient resource allocation, the consumption of fuels should be priced at long-run marginal cost (LRMC), i.e. the additional costs arising (saved) when consumers increase (decrease) their future demand for a given fuel. Competitive fuel markets would establish this price basis in any case by the operation of Adam Smith's invisible hand. However, national fuel markets are often anything but competitive being usually the province of state-owned or partially regulated monopolies. International supplies of competing fuels may provide an opportunity cost basis of LRMC for pricing purposes, but European governments have often been reluctant to open up indigenous fuel supplies to international competition – coal being the outstanding example.

Moreover, national governments, taking the view that fuel consumption has special social and strategic associations, have usually superimposed on the criterion of efficient resource allocation any number of additional policy objectives. Aside from the protection of indigenous resources already mentioned, these objectives include the desire to use energy prices as tax instruments, or as weapons in the fight against inflation, or as a means of redistributing income and wealth, or meeting political objectives such as security of supply. It remains true, however, that the principle of LRMC pricing is a useful benchmark against which to measure alternative energy policy stances.

Secondly, a variety of policy-oriented institutions developed. Some, like the OECD-based IEA were initially set up to confront OPEC, others, like the EEC, came to see energy economics as a critical factor in determining their long-run objectives. Such institutions naturally tend to develop policy stances along with those of member states. The EEC, for example, flirted with the notion, popular amongst American commentators, of joining a monopsonistic cartel of energy consumers, taxing oil imports to speed up the collapse of OPEC and so force a reduction in world oil prices. The EEC, along with the IEA and individual states has developed an embryonic oil stockpile and an emergency allocation system, and has funded conservation and alternative energy projects as well as nuclear fusion research. In particular, the EC Commission has, amongst its energy policy objectives, strongly supported the notion of pricing for rational use of energy (RUE), and such RUE pricing clearly has a close affinity with the principle of LRMC pricing described above.

In 1985, the Commission produced an extremely detailed projection of energy demand and supply for consideration by the European Council. This projection, *Energy 2000*, is predicated on the assumption that RUE pricing will be followed rigorously by member states. It reflects the closest link to date between the EEC's energy policy objectives, and the required instruments of policy making for their achievement.

What is particularly noticeable is that, while official EEC pronouncements on energy policy often reflect the rather bland compromises and platitudes that might be expected from intergovernmental debate, the Commission, in its commentaries and interpretation of guidelines has once or twice taken a robust, efficient resource allocation position which emphasizes that ideas of marginal cost based RUE pricing and financial viability of the regulated or nationalized fuel supply industries. In the field of energy policy, there is therefore considerably more material for analysis in what the Commission proposes, than in what the Council or Community disposes.

In summary, five principal components of an analytical framework for examining energy in Europe emerge from the discussion above:

(a) macroeconomic policy response to oil shocks
(b) relations with OPEC, and the nature of import dependence
(c) conservation and pricing developments in fuel markets
(d) investment in supplies of European fuel import substitutes
(e) the nature of European policy-making institutions and their relations with member states.

Each of the first four components is treated in turn in chapters 2 to 5 of this book. At various points the fifth element is brought in to fill out the policy dimension and to reinforce the discussion of the policy options that were considered. Finally, the EC Commission's *Energy 2000* projection and its implications are used in chapter 6 as a basis for discussing the possible evolution of European fuel markets.

The EEC energy balance

Having set out an analytical framework for examining energy in Europe, it is necessary, now, to look at a statistical overview of fuel supply and demand in the EEC. This is done by using the energy balance table for 1983 shown in table 1.1. This provides for the ten member states of the EEC a picture of the availability and final consumption of different fuel types, using a common unit: millions of tonnes of oil equivalent (mtoe). Thus coal supplies are converted to mtoe by using the calorific equivalent that 1 mtoe gives approximately the same heat as 1.5 million tonnes of coal and so on. A more detailed breakdown by individual member states is given in the Appendix at the end of the book.

The table distinguishes four fuel types in the columns: coal, oil, gas

Table 1.1 Summary energy balance for the EEC EUR10, 1983 (million tonnes of oil equivalent)

Source	*Availability and consumption of fuels*				*Total*†
	Solid fuels	*Oil*	*Gas*	*Electricity**	
1. primary production	174.0	132.5	119.8	89.8	516.1
2. imports (+)	56.7	481.4	78.2	5.9	622.2
3. exports (–)	17.6	192.6	30.0	4.1	244.3
4. stock change	–0.9	17.0	–2.8		13.3
5. gross consumption	212.2	438.3	165.2	91.6	907.3
6. bunkers	—	22.3	—	—	22.3
7. inland consumption					885.0
8. fuel uses and conversion					233.0
9. (total) final consumption	55.3	351.2	147.2	98.3	652.0
10. industry	35.7	49.3	59.6	41.8	186.4
11. transport	—	152.8	0.3	2.4	155.6
12. residential and tertiary	18.2	101.3	79.0	54.1	252.6
13. non-energy use	1.3	47.8	8.3	—	57.4

Source: Eurostat.

* Electricity includes heat.

† Individual columns or row totals may not tally due to rounding.

and electricity, but these convenient labels need to be interpreted with care. For example, along the rows labelled primary production and imports, 'oil' means crude oil piped from the North Sea or shipped to European refineries, whereas in the row labelled transport, 'oil' now means refined motor spirit and aviation gasoline for use in engines; in the row labelled residential and tertiary, 'oil' means heating oil used in household and office central heating systems, and so on.

The table begins in row 1 with primary production: i.e. the amounts of crude oil, coal, natural gas and primary electricity produced in the EEC member states. In 1983, coal was still the main indigenously produced fuel, chiefly in the UK and Germany, but was being closely caught up by crude oil very largely from the North Sea (UK sector). Almost of equal importance was Europe's production of natural gas which arises in the UK North Sea sector, the offshore and onshore fields of the Netherlands and in smaller quantities around the other member states. Primary electricity consists of nuclear generation and hydroelectric schemes, and it has recently shown the largest growth rate of any of the indigenous fuels. (Added into the electricity column are the very small amounts of solar power and cogenerated heat from the waste cooling tower water of coal and nuclear power stations.)

The next two rows indicate the importance of international trade in energy for supplying the EEC's primary energy requirements. The international trade in coal represents a small volume of net imports from outside the EEC that falls far short of the potential economic levels of coal imports, and it indicates what remains of several decades of protection of Europe's relatively very expensive deep-mined coal production from the competitive impact of the lower world market coal price. The latter reflects the much cheaper marginal extraction and transportation cost of Australian, American and South African surface-mined coal (along with some subsidized Eastern European deep-mined production).

The dominant fuel in trade is of course crude oil shipped to the European refineries of the international oil majors, the smaller international oil companies and those national European companies set up to retain control of refined products, such as Elf and Agip.

Net imports of crude oil at 288.8 mtoe are the largest single energy source defined in the balance, and are the reason why the three oil shocks have been of fundamental importance to EEC members, and also why dependence on imported oil has long been regarded as the EEC's principal economic problem in the energy field. This has now partly been joined by natural gas, a fuel again showing noticeable net imports to keep up the market shares of the rapidly depleted North Sea gas.

Row 4 indicates storage of fuels and includes the publicly known stock holdings of oil by both government and private companies used to top up the emergency allocation for use in embargo situations.

Totalling rows 1 to 4 provides figures for gross consumption of primary fuels, and subtracting the oil in marine bunkers, row 6, yields a total known as inland consumption of primary fuels. This is the total of fuels in raw material form available to final consumers. They, however, want to use secondary energy, i.e. fuels in a form suitable for use in industrial processes, domestic appliances, and central heating and air conditioning systems. Row 8 therefore totals the energy used in the conversion of primary fuels to secondary energy–the only net gain being in electricity which now has the production of coal- and oil-fired thermal power stations added in to nuclear and hydro capacity.[2]

In terms of the upper part of this table, many commentators, including the EC Commission, highlight the ratios:

(a) share of oil in gross energy consumption:
438.3 : 907.3 = 48 per cent
(b) supply dependence on imported oil
(481.4 – 192.6) : 907.3 = 32 per cent
(c) share of imports in gross energy consumption:
(622.2 – 244.3) : 907.3 = 42 per cent.

The corresponding figures for 1973 before the first oil shock, were:

(a) 62 per cent
(b) 62 per cent
(c) 64 per cent

and these indicate not only how far the EEC had come in ten years in responding to the oil shocks by switching away from imports, particularly oil, but also how reliant it still is on imports of all fuels, but oil in particular. The Appendix emphasizes the uneven distribution of this import dependence amongst the member states.

Returning to row 8, it can be noted that this summarizes the activities of the main fuel supply industries: not only the petroleum refineries, but also the electricity generation, transmission and distribution networks, the coal transportation system, and the gathering and transmission system for natural gas. In one sense these are the primary focus of energy policy making. It is these industries which feel the first impact of changes in primary energy prices, and which pass these on in one way or another to the final consumers. In Europe they are frequently state-owned enterprises, and even in Germany where the tradition of private sector cartels remains, the principal electricity authorities are very largely owned by federal and state governments and municipality shareholders. These fuel industries share, therefore, several characteristics:

1. They are responsible for the large-scale investment decisions in new fuel supplies: e.g. nuclear power, natural gas purchase contracts and so on.
2. They are usually 'natural monopolies' because of the importance of economies of scale in production, and, particularly, distribution, and are therefore either wholly owned or closely regulated by government.
3. They are often used as instruments of tax gathering or macro-economic policy since their products are so widely consumed, their investment programmes so large, and their pricing can be readily manipulated by non-competitive market forces.

After conversion from primary energy, the table illustrates final consumption of fuels as used by three main market sectors: industry, transport and residential–tertiary; this last comprises households as well as public and commercial offices, schools, hospitals and so on, and in terms of energy consumed is the largest sector.

Each sector has its own preferred mix of fuels according to the end uses to which energy is directed. Industry will require energy for steam raising, industrial processes, heating of factories, industrial cleaning, electrical motors and controls. It has an almost equal balance of all four fuel types. The transport sector uses, almost exclusively, oil products, while the residential–tertiary sector has most noticeably switched away from coal over the post-war period (household use of coal being apparently a statistical example of what economists refer to as an inferior commodity, i.e. its use has declined with rising living standards). This sector now uses electricity, oil and gas for space and water heating, along with electricity and gas for cooking. Finally, a fourth sector: non-energy use refers to the market for fuels as petrochemicals, feedstocks, lubricants and fertilizers, where the end use does not involve the provision of energy in the form of heat.

The fuel market impact of oil price shocks is finally transmitted through the prices charged by the fuel industries in these markets, and analysis of the impact of price and income changes depends on measuring the three principal economic determinants of energy demand arising from market behaviour in these sectors:

(a) the force of technological progress in designing fuel-using appliances
(b) the responsiveness of consumer and producer demand for fuels to price changes (price elasticity of demand)
(c) the effect on demand for fuels of both the long-term growth of national income and short-term fluctuations in it (income elasticity of demand).

Among their other influences, the three oil shocks caused economists very greatly to improve and enhance their measurement and understanding of these three factors.

The fuel supply industries themselves will behave in ways characterized by these elasticities in order to achieve their own or their government-regulated objectives, and the most important and responsive is the electricity supply industry.

Table 1.2 casts additional light on electricity supply in the EEC by looking (for EUR10) at fuel mix in electricity generation and its changes over the period 1973–83.

The electricity utilities in the EEC, although overlaid with other objectives, will at least seek to minimize the cost of generation by changing the mix of fuel types and generating plant used. Table 1.2 gives such a breakdown of the percentage shares of the different fuels. In 1973 input of oil products accounted for 32 per cent of the kilowatt hours (kWh) of net electricity generated, but by 1983 the share of oil products had dropped to 13 per cent of generation input. Since this shift from oil was accommodated within a 25 per cent growth of electricity production over the period it actually represented a 49 per cent fall in the level of oil input (see column 3 of table 1.2). Natural and derived gas were also reduced as electricity inputs as a consequence of the attempt to concentrate North Sea and Dutch gas on premium uses rather than as a substitute for oil burning. (The overall use of natural gas in other sectors showed a net increase like electricity.) The other three inputs – solid fuels, nuclear, and hydrogeothermal – were substitutes for gas and oil, with the last of the three maintaining its share within the overall electricity expansion. Electricity generation switched largely towards solid fuels and nuclear in line with EEC energy policy objectives, with nuclear showing by far the largest expansion.

Table 1.2 Electricity generation in the EEC EUR10, 1973–83

Fuel input to generation	*Share of net kWh generated (%)*		*Change in kWh generated (1973–83) (%)*
	1973	*1983*	
solid fuels	38	43	+40
oil products	32	13	−49
natural and derived gas	12	9	−10
nuclear	5	22	+414
hydrogeothermal	13	13	+31
total	100	100	+25

Source: Eurostat.

The nuclear emphasis has not been widely balanced – France and Belgium have accounted for the lion's share of this growth, along with Germany and the UK, while Italy and the Netherlands have maintained a minimal nuclear share. The Netherlands and Denmark were both found to be undecided about what role, if any, nuclear energy should play in electricity generation. By 1983, the EEC nuclear share, at 22 per cent, exceeded those in the USA (12.6 per cent) and Japan (18 per cent).

This picture of energy sources and use in the EEC is the 1983 snapshot – one decade after what has been thought of as an energy crisis, (though mistakenly, as argued above). But the fact that it is ten years after the initial oil price rises should not be given undue emphasis. The course of oil prices has consisted of extremely volatile episodes around a long-term trend, and it is only an artificial device to select specific dates in relation to one or other period of oil market disturbance. Nevertheless, the idea of 'three oil shocks' provides a convenient peg on which to hang the overview of energy in Europe. The nature of these shocks can now be considered in a little more detail.

The historical background

After the Second World War, the economy of western Europe was largely self-sufficient in energy production, and over 80 per cent of the primary energy requirements of the countries now forming EUR10 were met by indigenously produced hard coal and lignite. In the following thirty years two trends were outstanding. On the one hand there was an unprecedented expansion in income levels and, coupled with this, energy consumption. Primary energy consumption for western Europe as a whole rose by about 170 per cent between 1950 and 1973 (an annual rate of increase of over 4 per cent). On the other hand, indigenous production was virtually stagnant between 1950 and 1973, with an annual rate of increase for western Europe of just over half of 1 per cent (Ray 1982).

The huge substitution implied by these figures was away from indigenous coal and towards imported oil: coal use in western Europe declined from 86 per cent of consumption in 1950 to 25 per cent in 1973, while oil use expanded from 12 to 59 per cent (Ray 1982, EC Commission 1978). Throughout the period the underlying preoccupation of both national governments and the infant Community institutions was with future energy shortages, since it was very common to use forecasts that allowed for continued demand escalation without a corresponding price response.

In 1973 the Commission's forecast for 1985 for gross energy consumption for *nine* member states was about 1800 mtoe, an overestimate of almost 100 per cent on the actual turnout of 965 mtoe for *ten* member states.

The long-term impact of this switch from coal to oil can be seen in three developments. Firstly, a distinction that had existed in 1950 between Europe's energy producers (Germany, France, Belgium, the Netherlands and the UK), and its energy consumers (Denmark, Greece, Italy, Ireland and Luxembourg) gradually disappeared so that by 1973 all member states of EUR10 were net energy importers.

Secondly, the coal to oil substitution had a regional impact since, immediately after the Second World War, industry was still largely concentrated in the coal-producing central areas of northwestern Europe. However, the logistics and transport costs involved in a switch to crude oil imports led to industrial development along the coastal regions and near the main European seaports. Pipelining crude oil to the old established industrial regions was less economic than relocating near oil landing terminals and refineries. This had the effect of permitting industrial development in parts of southern Europe, and led to reduced economic activity in the old coal-producing areas. Coupled with this relocation, the erosion of the economic attractiveness of coal as an energy input led to further unemployment and declining economic activity in the traditional areas of heavy industry.

This was offset up to 1973 by the general rise in European prosperity, but became a critical problem after 1973 when, as chapter 2 indicates in detail, oil price rises pushed Europe into recession. The cheap oil growth era, partly the result, as chapter 3 shows, of the American decision to restrict its own oil imports, had the effect of permitting both Europe and Japan to rival American economic growth rates and prosperity over the 1960s.

A third development associated with the shift to oil was an enormous expansion in European refining capacity to cope with the increased demand for petroleum products. This was the province of both the long-established major international oil companies as well as the growing numbers of national European oil companies, all of which contributed to the problems of regulation and competition monitoring for the young EEC. By the time of the oil crises of the 1970s this refining capacity was massively excessive, and was subsequently being undercut as the Middle Eastern producer states set up their own subsidized refinery capacity, partly as a means of monitoring (and evading) their own cartel arrangements. This led eventually to large-scale rationalization of refinery capacity, though there has always been the problem that countries need some basic refinery capacity since access to alternative crude oil supplies in a crisis is negated if there are no facilities for converting the oil to petroleum products.

This was the energy context in which the Community institutions took shape. As is well known, the initial vision of a united Europe evolved in the European Coal and Steel Community (ECSC) whose primary task was the dismantling of the Ruhr coal cartels in which it was supported by both French and German governments (George 1985), one to secure cheap supplies, and the other to exhibit its free market orientation along with membership of an international forum.

In 1958–9 the High Authority of the ECSC faced and, by common consent, failed its first major test. As oil imports started to rise indigenous coal stocks in the ECSC quadrupled, and the coal-producing member states imposed coal import quotas, threatening the idea of a common market. When the High Authority sought to declare a crisis and impose production as well as import quotas, it failed to achieve member states' agreement, chiefly, it would appear, because of reluctance to transfer policy making away from national governments.

Fuel sources other than coal came under the aegis of the EEC and Euratom. Nevertheless, as the Commission itself points out (EC Commission 1985c) the concept of energy policy as such is not mentioned in any of the three Treaties, although all three relate to the energy sector. To some extent the omission was recognized from the outset and the member states at the 1956 Messina Conference had instructed the High Authority of the ECSC to consider a coordinated energy policy.

Following the establishment of the EEC and Euratom, this initial development from the High Authority led eventually in 1962 to a combined Working Party on energy forecasts, data collection, price monitoring and so on. Despite a lack of agreement on energy policy, the first energy policy instrument, the Protocol, was signed in 1964. Generally this called for a harmonization of national policies in preparation for a common energy market, for state aids in the rationalization of the coal industry alongside a programme of Community aid, for a common policy of diversifying oil and gas supplies and for supporting nuclear research.

Throughout the period the rundown of coal was a preoccupation of member states and the Community. The Commission has stated that its policy tried to ensure 'that coal should not be priced out of the energy market and so high price levels for other sources of energy, and in particular those of imported energy, especially oil, were maintained so as to protect indigenous coal, and allow it to compete favourably' (EC Commission 1978). The specific aids to coal were in addition to this policy stance. It is tempting to see the subsidies to coal as an indicator of a low-cost energy strategy especially when during the 1960s more competitive markets for imported oil were permitted. However, it is arguable (see chapter 3) that the policy of coal protection was so endemic to the large European producers that the low fuel prices of the 1960s occurred very much in spite of rather than because of member state or Community action. The low oil price era indicated an inability to control market forces rather than a willingness to adapt to them, and this political commitment to high energy prices has been offered as one reason why OPEC was subsequently so successful at raising oil prices due to temporary shortages when long-term supplies were abundant.

The next step in Community energy policy was the 1968 'First Guidelines for a Community Energy Policy' and the main issues raised were the relations between energy importers and exporters, the regulation and monitoring of the Community's oil companies, the development of nuclear power and the future role of coal. No immediate pressing problem indicated any particular policy development – other than the idea of a common energy market. The overall aim of improving competition in fuel supply was qualified by the suggestion of coal import quotas, and there was some support for an embryonic oil stockpiling programme of 65 days consumption equivalent. The 'First Guidelines' received little government support, other than for the most basic principles that policy making remained a national responsibility and could include state aid to the coal industries, and Black (1977) comments that 'little in the way of concrete output emerged between the Guidelines and the 1973 oil crises'.

A great many commentators offered, at the time, as the principal argument for a common energy policy a view expressed by the Commission in the words: 'creation of a common energy policy is a fundamental factor in the creation of European unity' (EC Commission 1978). In other words, a form of political idealism was the only basis suggested for policy action, against which there was the massive obstacle of the pursuit of national self interest. As chapters 2 and 4 will show in detail, policy ideas only become concrete when an actual threat to living standards from the use of OPEC's monopoly power emerged, and, as will become apparent, since that threat had a macroeconomic (rather than an energy policy) dimension, it is not surprising that the policies that evolved immediately after 1973 were macroeconomic in nature. The only real cause for surprise is that it could ever have been imagined that national self interest would not be dominant until an actual threat that was amenable to concerted action emerged. As chapter 3 will show, there is considerable evidence, to support the view that in the event of a probable threat, changed risk perceptions on the part of market participants will still ensure that *individual* risk avoiding behaviour under market pressures is the first line of defence.

This review of the context of European energy developments takes the situation up to 1973, the occasion of the first oil shock. The principal argument of this book is that energy only became a primary political problem under the impact of the shock to the European macroeconomic system from oil price rises, and that, as a consequence, a study of energy in Europe has to start seriously with a consideration of the external stimulus of the oil shocks.

Under this stimulus, it will always be the case that national responses will dominate while at the Community level the most that can be achieved is the articulation of objectives and the monitoring and analysis of fuel pricing and investment policy. One exception to this might arise in the context of import supply security if it could be shown that the Community as a monopsonistic group can wield more significant market power than as a collection of individual importers. The debate on this issue and the related idea of an energy import tax is considered in chapter 3. The articulation of demand targets and pricing policy is considered in chapter 4, and the economic issues surrounding individual member states' fuel supply policies are examined in chapter 5.

Three oil shocks

The 'OPEC decade' is used as a convenient starting-off point for discussion of the energy in Europe theme, and, rather artificially, the period since 1970 is here said to include three oil shocks: two crude oil price rises, and one, more recent, period of price erosion.

Figure 1.1 traces the history of the price of crude oil set by OPEC both in nominal, i.e. $ per barrel, terms and in real terms as measured by deflating the $ per barrel price by an index of the export price of manufactured goods for the OECD.[3] This deflator measures, approximately, the ability of oil exporters to purchase the manufactured goods traded across the world, and the necessary export values that world manufacturers must achieve to pay for their imports of crude oil.

After a long period of steady, indeed slowly declining oil prices, up to 1970, the first stirrings of OPEC market power emerged with the insistence by the host countries around the Persian Gulf and North Africa that the oil companies should pay higher royalties and tax on their concessions. Gradually, through a process of nationalization, raising of the royalty rates and finally straightforward announcement of crude oil prices, the host countries, working together as OPEC, were able to capture an unprecedented degree of monopoly power in the world oil market. From a price of $1.39 per barrel for Arabian light crude prevailing on January 1970, prices rose, in the first oil shock of 1973–4, to a level of $10.46 per barrel by 11 January 1974. The single largest jump of over 130 per cent occurred between December 1973 and January 1974 during the Arab–Israeli war. Between 1 January 1973 and 1 January 1975 the price rose by over 475 per cent.

After some stability in the nominal price (and a fall in real terms), the second oil shock coincided with the Iranian revolution of 1978–9. Between 1 January 1979 and 10 January 1981, the price rose from $14.55 per barrel to $34.00 per barrel, a rise of 134 per cent. At this

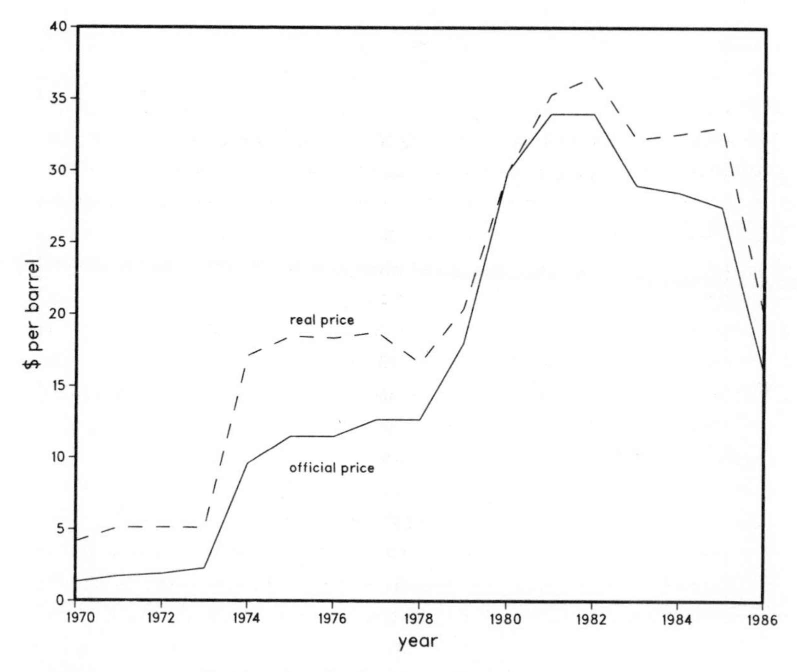

Figure 1.1 Official and real oil prices, 1970–86
Source: Exxon, OECD

time higher quality crude from the North Sea reached $41 per barrel in following prices on the Rotterdam spot market.

What is often referred to as the third oil shock was really the market response to these price rises. Demand for OPEC oil slumped and the more expensive fields of the North Sea became economic. With the resulting excess supply situation, OPEC found itself compelled to acquiesce in lower crude oil prices, even with production cutbacks. By December 1984 the price was back at $29 per barrel, but below this on spot sales, and by 1986 had fallen to a nominal price of $16 per barrel.

For convenience, these three periods of shifting prices are centred on 1973–4, 1979–80 and 1983–5 in looking at the response in European energy markets (figure 1.2).

Figure 1.2 traces the trend in primary production (row 1 of table 1.1) and net imports (the difference between rows 2 and 3 in table 1.1) for the EEC (EUR10) over the period 1970–85. There are both

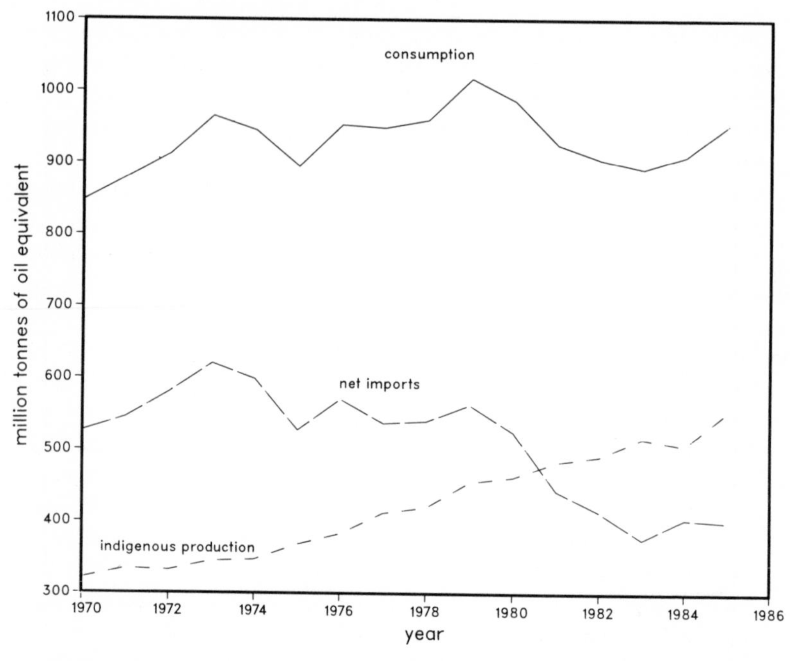

Figure 1.2 EEC energy consumption, production and imports, EUR 10, 1970–85

Source: Eurostat

similarities and differences between the price change episodes, but several important characteristics can be noted.

To begin with, the period before the first oil shock is characterized by the EEC's enormous dependence on imported energy with net imports being more than one and a half times as large as the EEC's own primary production. The shock of the large price rises and the consequent transfer of real income from the EEC (and the rest of the OECD as well as the nodc's) could have produced an immediate and prolonged slump. In fact it is clear from the figure that by 1975 imports had recovered to their pre-shock levels suggesting that national income levels had held up rather well. Indigenous production however began a faster upward trend which continued unbroken to 1984.

The first oil shock appears therefore to have been relatively short-lived, despite giving rise to fears of an energy crisis. The second oil shock of 1979–80, by which time EEC energy imports were at a record high level, as was energy consumption, had by contrast a much longer and more serious recession following it. Clearly there must have been policy effects that did not hold up the levels of national income following the second shock. In addition, although the price rises were smaller in percentage terms, because they started from a higher price level, economics would predict a stronger price response (higher price elasticity) to drive down imports.

The prolonged recession and fall in imports was nevertheless accompanied by steadily rising indigenous energy supply until 1983–4. Then, energy production appears to have fallen for the first time in fifteen years and imports rose. However, the rise in imports was partly due to the recovery in national incomes as oil prices fell in the third oil shock, and it did not involve greater use of OPEC oil. Moreover, it was accommodated within a lower overall energy consumption than was present in the second oil shock. The fall in indigenous production was wholly accounted for by the prolonged miners' strike in the UK, and reflected non-availability of high cost coal supplies rather than a reduction of the EEC's long-term ability to produce energy. By 1985, the temporary reversal had been eliminated, imports were again falling and indigenous production rising. This was the context in which the OPEC cartel fell into serious disarray and the oil price dropped by about 50 per cent as the 1986 OPEC meetings failed to agree on anything other than the fact that each member was going to raise its share of the world oil market. By 1986, the real price of crude oil to the European economies was only about 20 per cent higher than it had been at the end of 1973.

2 The oil shocks

Fundamental questions

The oil price rises of 1973–4 and 1979–80 have not lacked forceful descriptions: e.g. 'amongst the largest macroeconomic impulses the OECD has had to analyse over the last fifteen years or more' (Llewellyn 1983), or 'a milestone in modern energy history' (Brondel and Morton 1977). The third shock, smaller in size, more delayed, and in the direction of weakening oil prices may also turn out to be of great long-term importance.

Analysis of the oil shocks is fundamental for several reasons, the most important of which is that they are the proximate cause of the view that energy supply poses a problem for Europe and the EEC in particular. By examining the impact of oil price changes, it should be possible to see why 'energy in Europe' has become an important focus for economic and political initiatives, and it may also be possible to see why the idea of a Common Energy Policy (CEP) gained little if any ground as a way of responding to the energy supply problem.

Several reasons why energy supply should be regarded as critical for the EEC have been suggested. Amongst the most obvious and superficial is that fuels, particularly oil, were approaching severe scarcity if not exhaustion in the 1970s. There is (and was) no credible evidence for this as the weakness in oil markets that emerged in 1982 showed clearly. A second reason offered was that energy was too important a commodity to be left to unregulated market forces and that by its very nature it required a policy stance. There is little merit in this argument, confusing, as it does, the fundamental importance of energy as a concept in physics, and the actual buying and selling of particular barrels of oil or tonnes of coal. It is quite clear that individual fuel markets do exhibit the usual demand and supply responses of market behaviour (though, clearly, some types of fuel supply like electricity distribution tend naturally to monopoly market structures).

A third reason why energy supply attracted policy makers' attentions might be that the institutions of different EEC national governments led naturally to intervention in and organization of fuel markets, so that a supranational preoccupation with energy supply for the EEC automatically followed. This can partly be seen in the early formation of the European Coal and Steel Community (ECSC).[1]

However, it is clear that preoccupation with a European-wide energy policy was not widespread until after the first oil shock, so that it is necessary to look for some factors other than the existing nature of energy policy institutions, both nationally and at the EEC level, for the stimulus.

What is clear is that the first two oil shocks produced the largest shift in the terms of trade against the EEC (and the rest of the OECD) experienced in the post-war period. These massive redistributions of resources from western Europe, Japan and the USA to the OPEC producers necessitated fundamental structural readjustments in all the oil importing economies including the non-oil developing countries (nodc's). There is considerable evidence to support the view that the adjustments were initially misunderstood and badly managed, and hence exacerbated by government policies. It is the resulting income losses and social frictions in market adjustments that can be said to have made energy in Europe a policy problem. But clearly, energy supply is only a vehicle here for the real income losses involved.

In the long run different national economies began to adjust their production structures differently depending on their ability to produce oil import substitutes, while in the short run the important questions concerned national and international policy to deal with real income shocks. In other words, though energy in Europe is convenient shorthand for a whole spectrum of issues requiring analysis, energy supply only rose to policy prominence because it was the channel through which western Europe suffered major short-term income losses because of the large share taken by imported oil in its energy usage.

This possibly throws light on the absence of a CEP: the income adjustments required *macroeconomic* and not energy policy responses in the short run. In the long run, apart from the common necessity of not obscuring market price signals, and minimizing the impact costs of oil embargo, structural adjustments necessarily *differed* amongst the EEC member states. In facing short-run income shocks, a CEP was irrelevant, and in adjusting to long-run structural changes, common policy responses are not necessary, likely, or desirable.

In the analysis which follows three broad country groupings are distinguished as the chief entities involved in the oil shocks. The initiators are the Organization of Petroleum Exporting Countries (OPEC), while the recipients are the nodc's and the industrialized countries. This last group is largely represented by the Organisation for Economic Cooperation and Development (OECD), and the countries of Europe as a whole, as well as the EEC, are part of this group. Within the OECD a specific group, the major 7 economies, are frequently used to illustrate policy responses and adjustments. This subgroup comprises the USA, Canada, Japan and the four major economies of the EEC: West Germany, France, Italy and the UK. These countries are also prominent because of the series of world economic summits which they have organized over the 1970s and 1980s. In measuring the impact of the oil shocks on the EEC in particular, and Europe in general, it is therefore both useful and convenient at this stage to concentrate on the OECD and the OECD major 7 as the representative groups.

Long-run impacts of oil shocks

The critical overall measure of OPEC's decision to wield monopoly power in the world oil markets is the graph of the real price of oil illustrated in figure 1.1 in the previous chapter. This converts OPEC's official oil price into a measure of its real purchasing power in the developed countries who form OPEC's principal customers. The resulting picture gives the terms of trade facing the OECD in using its exportable industrial products to buy a major importable commodity, crude oil. The first and second oil shocks show how the terms of trade shifted against the OECD in two massive jumps of approximately 300 and 150 per cent. The third shock shows the oil price set by OPEC falling but with the OECD's prices falling faster, even the 1982–4 period ended with a minor worsening in the OECD's terms of trade.

The first two oil shocks are, on each occasion, estimated to have transferred to OPEC from the OECD and the nodc's an amount of real income equivalent to about 2 per cent of OECD GDP. By mid 1983, the third episode appeared to have transferred back about half of one per cent of OECD GDP.[2] These are the fundamental income shifts referred to collectively as the oil shocks.

The impact of the first shift has to be seen in the context of the build-up to it. Prior to 1973 several important trends were already observable in the world economy. Firstly, the fixed exchange rate system set up in 1944 at Bretton Woods, and which gave birth to the International Monetary Fund (IMF) and the World Bank, had effectively broken up. This system was designed to permit countries to maintain stable exchange rates even through periods of balance of payments surplus and deficit by arrangements to lend and borrow between the Central Banks of the major economies. However, some countries, notably West Germany, which had wished to follow economic policies embodying more restrictive monetary policy than their trading partners found that the system offered very large gains to foreign exchange speculators at little risk. Exchange rates were held fixed until the last moment when countries in surplus were faced with the choice of huge monetary expansions to catch up with their partners or large appreciations of their currency. Those in deficit faced the eventual dilemma of large monetary contractions or depreciations. By the early 1970s the world trading system was using flexible exchange rates which allowed national governments to follow their own monetary policy without having to worry about accumulated speculative pressure on their exchange rate. At the same time the system of Central Bank lending and borrowing was gradually being overtaken by the large-scale lending programmes of private sector banks.

A second long-term trend was the structural shift in world trading patterns. As the less developed countries gradually industrialized, the world's supply of manufactured goods had been expanding and its prices falling relatively to the prices of services. Services, which are often concentrated among the non-traded commodities in the economy include wholesale and retail trade, transport and communication, insurance, banking consultancy and R & D, and educational, medical and entertainment services. Many of the OECD countries had been approaching the sort of post-industrial state in which this sector grew relative to industry and agriculture. Figure 2.1 shows, for Europe as a whole, the changes in sectoral value added as percentages of gross domestic product (GDP) for the ten years preceding the first oil shock and the nine years following it. In both periods the share of agriculture in output declined steadily, and prior to 1973 the share of industrial output had also been declining while that of services had grown substantially.

Superimposed on the structural changes in trade and the erosion of the international financial system, was a long-running synchronized boom for the OECD countries, and, associated with it, was a worldwide inflation that had started to take off in the late sixties, fuelled,

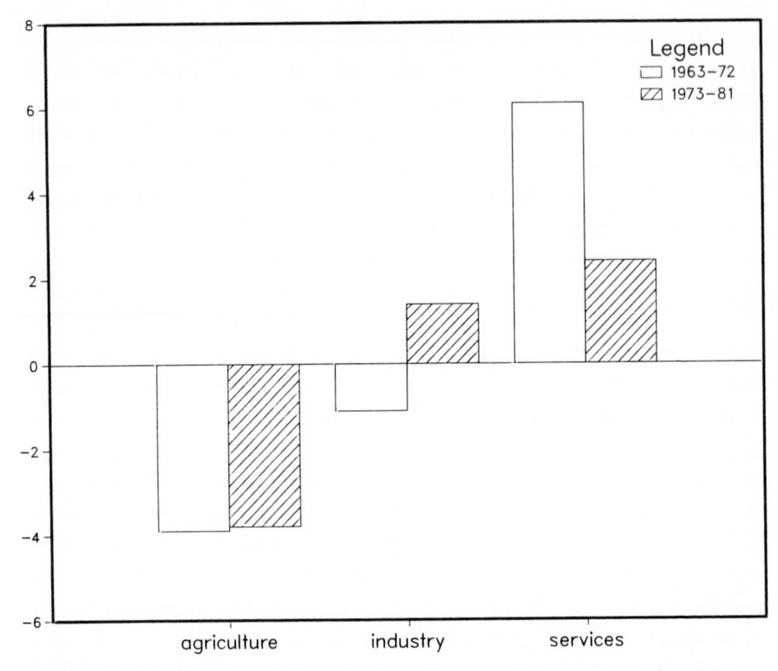

Figure 2.1 Europe: changes in sector value added as per cent of GDP
Source: IMF

some observers believe, by the monetary expansion of the USA in financing the Vietnam War, and spread through the fixed exchange rate system.

Figure 2.2, which illustrates inflation rates over the whole post-war period, shows clearly the take-off into inflation in the 1960s and early 1970s. This worldwide demand expansion, which included a primary commodity price explosion, was one of the factors which permitted OPEC to have such success in pushing up the price of oil, since, in the short run, the oil importers had a very limited ability to cut demand quickly in response to price rises.

The first two oil shocks had an enormous impact in disturbing the evolution of the world economic system for both the industrialized countries of the EEC and OECD, and the nodc's. The immediate macroeconomic consequences are considered later, but attention can first be focused on the long-term responses of the oil importers.

In the absence of offsetting government policy, the long-term

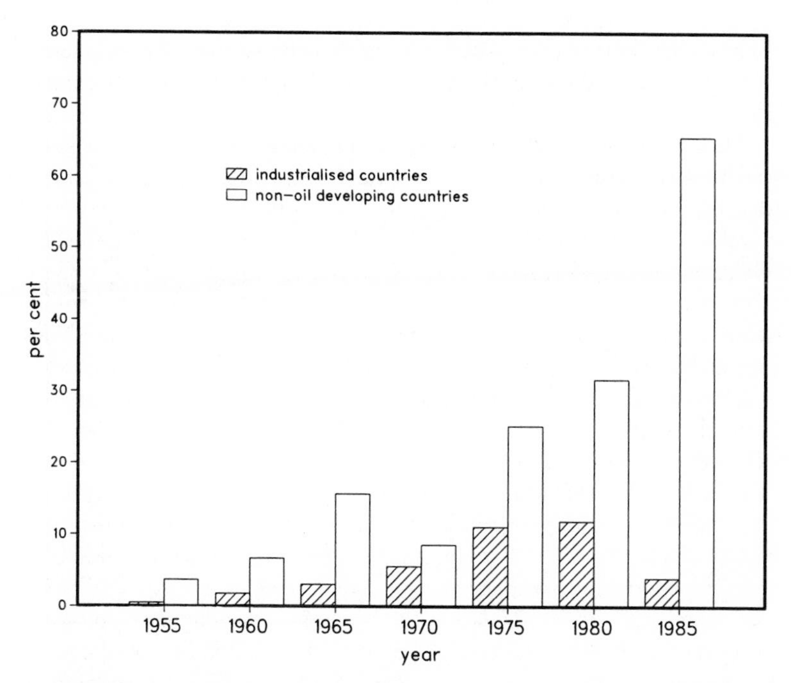

Figure 2.2 Inflation. Rates of change in consumer prices, 1955–85
Source: IMF

responses to a terms of trade shift against the oil importers can be viewed under three headings.,

First, there is clearly a loss of real income as importable goods rise in price and exports fall. This need not be permanent because on the one hand the oil exporters have experienced a rise in real income which they will eventually want to spend, and on the other the oil importers still have the opportunity to re-establish their real income levels by structural and technological change. Nevertheless the real income loss may persist for some time, and in the case of the oil shocks it was severe and persistent enough to produce the worst recession since the 1930s in the developed countries.

Secondly, there is a substitution in consumption away from importables and towards consumption of exportables, so that the oil importers can be expected to reduce their consumption of imported oil quite heavily over the long run. This in itself makes them less vulnerable to repeated use of oil monopoly power. Again this is likely to be a long delayed process and it is widely believed that oil consumers were still responding to the first oil shock when the second came along.

Thirdly, there is a substitution in production away from exportables and towards import substitutes. Figure 1.2 has already made it clear that the EEC has gradually been extending its indigenous energy production over the period of the 1970s. This production response may eventually lead to a reversal of all or part of the initial terms of trade deterioration as the supply of importables and import substitutes rises while their demand falls, and meanwhile the supply of the OECD's exportables falls just as OPEC's real income is expanding. Certainly it was the case that the first and second oil shocks contributed very largely to the commercialization of the North Sea oil and gas reserves of Norway and the UK.

However, it is true that not every oil-importing country can expand its production of import substitutes. Those that can, the industrial energy producers, will do so and, in the process, squeeze out production of exportables like manufactures which they previously used to pay for oil imports. In this sense the possession of indigenous energy resources which have been revalued upwards by the first two oil shocks has altered the comparative trade advantage of such industrialized producers. On the other hand, oil importers without large recourse to indigenous energy import substitutes will have to pay for their higher priced imports by a much more rapid contraction of import volumes together with a continuing or recharged export drive.

Looking at figure 2.1 again, it can be seen that in the 1973–81 period, industrial output, the initial basis of Europe's exportables, recovered part of its share of GDP while services' growth faltered. This hiccup to the long-term structural changes observed before 1973 occurred because Europe, as a whole, had to expand its base of manufactured exportable goods to meet its new higher import prices. Within this adjustment, the European energy producers, the Netherlands, Norway and the UK switched away from being net exporters of manufactures towards production and export of imported oil substitutes. Over the 1973–9 period both Norway and the Netherlands had deficits on the manufacturing trade balance with surpluses on the trade balance for other goods including energy. The UK reached this position in 1983.

In contrast, the oil importers without energy substitutes continued to import energy while maintaining their positions as net exporters of manufactures: over 1973–9 the manufacturing trade surpluses as percentages of total imports for France, Italy and West Germany were, respectively, 9, 25 and 40 per cent.[3]

As these long-term adjustments evolve, the absolute level of oil imports falls until trade approaches a balanced position once more. Depending on the size of the relative price changes, this balance may involve a smaller share of exports in production so that structural changes amongst the OECD trading partners may be quite fundamental. Except in the short run aftermath, however, there is no reason to believe that the terms of trade shift will automatically mean prolonged trade deficits on the current account. Nevertheless, this was one characteristic of the adjustment to oil shocks, along with a deep and prolonged recession in the EEC.

However, this description of the long-term responses, the implications of which are made clear in the discussion of EEC energy market developments in subsequent chapters, must be qualified by examining the short-run macroeconomic consequences of the oil shocks, and by noting that government policy played an active role in these.

Short-run impacts of oil shocks

The short-run impacts of the OPEC oil price rises, which can be said to have lasted for two to three years in each case, are the factors which brought home to people the problems of energy supply. The real income losses and market upheavals really constitute what came to be called the 'energy crisis'.

It has already been noted that each of the first two shocks transferred approximately 2 per cent of OECD GDP away from the oil importers and to OPEC. This immediate loss of real income might only have lasted for a short time had OPEC rapidly respent its income gains. For several reasons this did not happen and there was therefore a multiplied drop in real income throughout the developed world. At the same time inflation in the oil importing countries was given a boost by the rise in industrial costs associated with the oil price rises.

The resulting 'stagflation' can be seen in figure 2.3 which shows, for the total OECD area, the fall in real growth rates over the periods 1973–5 and 1979–81, and the increase in inflation rates over the periods 1973–4 and 1979–80. Overall growth suffered and inflation escalated. Nevertheless, although the size of the additional oil import bill for the OECD as a percentage of gross national product (GNP)

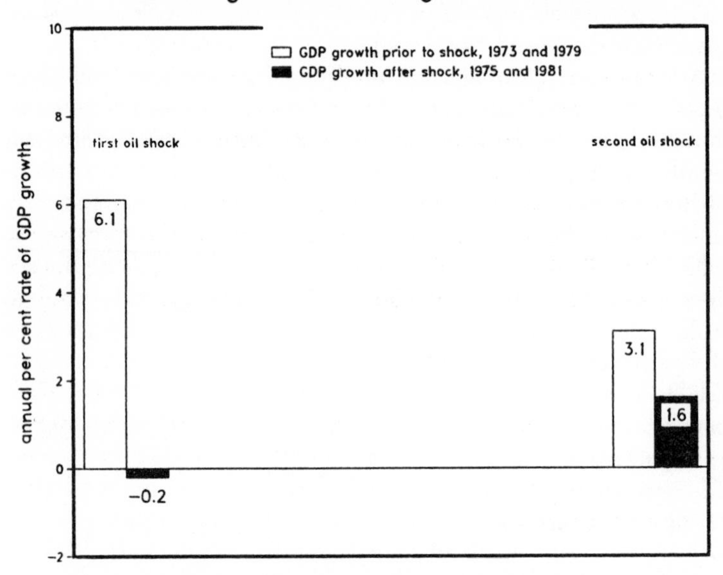

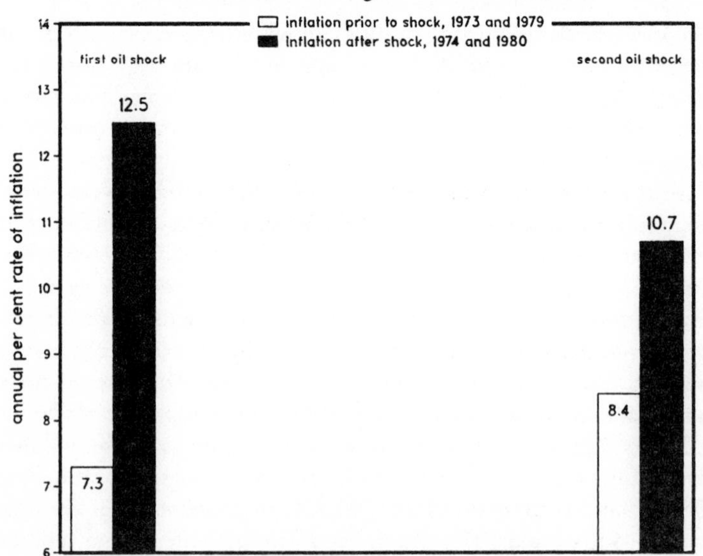

Figure 2.3 Oil shock impact on growth and inflation
Source: OECD

was approximately the same in both cases, the first two shocks did exhibit considerable differences. The 1973–5 recession into negative growth was the worst since the 1930s while the inflation rate almost doubled. In the second oil shock growth contracted from an already low level to a lower but still positive rate, and the control of inflation which had commenced after 1975 was largely maintained as the oil price impact on all prices was much less.

Before considering in detail what in fact happened in the oil shocks, it might be helpful to set out the main trends which economics would predict if there were no policy responses, or other external factors affecting events. In other words to describe briefly a 'model' of an oil shock.[4]

The starting point is a large transfer of income taking place from the OECD to OPEC. As long as the gainers of this additional income do not spend it as rapidly as the losers reduce their spending, *the first impact of an oil shock must be a fall in aggregate demand for OECD production.*

This would show up as lower real GNP in the OECD and a tendency for prices to fall.

At the same time, however, since oil is a net import to the OECD to be used as an input in the production of GNP, there must be both a short-term rise in the costs of producing GNP, and a long-term fall in the potential level of GNP which the OECD can produce. The latter factor reinforces the actual fall in GNP arising from the aggregate demand effect considered initially, while the former factor offsets the tendency for prices to fall and may indeed push them up above the pre-shock equilibrium.

The second impact must therefore be higher prices and a further fall in GNP arising from *reduced actual and potential aggregate supply.*

Following these short-term impacts there is a variety of possible medium- and long-term effects.

In the medium term, the OECD economies find they are going through a prolonged oil-induced recession with accompanying unemployment. Left to itself, actual GNP may start to rise as the recession pushes down prices and re-establishes consumers' real spending power; the OECD economy may move once again towards a long-run equilibrium with stable prices and low expectations of inflation. Nevertheless, its potential GNP will still be below that of the pre-shock situation because a critical input to production, oil, has risen in cost.

It is only with the longer term adjustments noted earlier that this can be overcome as producers and consumers conserve energy and switch to energy-saving means of production. Potential GNP therefore approaches its pre-shock level at a rate that depends on long-run fuel substitution, conservation, technological progress and investment in new capital equipment.

Within this overall picture, two complications need to be borne in mind. Firstly, the overall effects conceal differences in the impact on individual OECD members. In some the demand factors may dominate, in others it is the supply–cost factors which will be important. Secondly, some types of markets in the OECD economies – notably financial and currency markets – are known to respond much more quickly to outside shocks than others, such as the labour market. This divergence in speed of response is believed to cause some prices in the economy (e.g. interest rates and exchange rates) to 'overshoot' their new equilibrium positions before returning to equilibrium, with the consequence that cyclical movements are induced in the economy's return to equilibrium.

Turning from this description of theory to the actual circumstances of the oil shocks, it is clear, however, that the fundamental problems of stagnation and inflation are the result of not only OPEC's initial boost to world oil prices, but also the policy responses, both direct and indirect, of the national governments of the developed oil importers and nodc's. The fiscal and monetary stances of these governments then partly show up in two further problem areas said to arise from the oil shocks: the growth in government budget deficits and the size of current trade account deficits.

How then did the initial impacts shown in figure 2.3 become amplified and later modified in reality by private and public sector responses?

Consider first the real income transfer. This had an immediate and sustained effect in diminishing demand for OECD exports because the OPEC members did not respond to their real income gains by increasing their spending in the same proportion. Figure 2.4 indicates how, following each of the oil shocks, OPEC at first accumulated its real income gains in the form of current account surpluses. Several reasons have been offered for this. The OPEC nations have been categorized into two groups of 'low' and 'high' absorbers depending on their ability to increase their annual growth of import volumes. Among the low absorbers were the small, sparsely populated but superlatively oil-rich Middle Eastern countries like Kuwait and Saudi Arabia, who it is

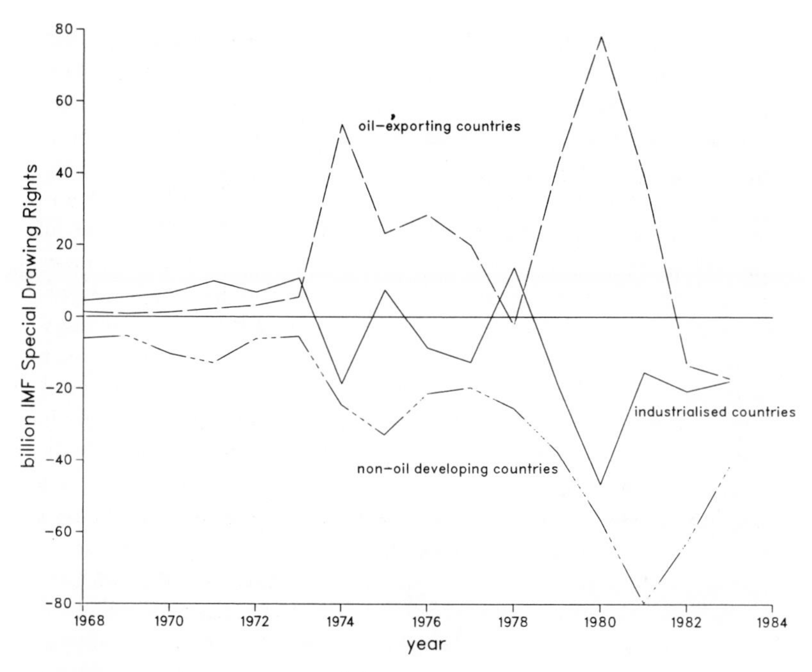

Figure 2.4 Current account balances, 1968–83
Source: IMF

generally held, could increase imports by at most 20 per cent per year after the first oil shock. This is because excessive imports were perceived as wasteful and therefore restricted by official policy, because there were limited port and entry facilities, and because respending the new oil revenues required prior planning and organization. There is also the suggestion that if the price rises were only to prove temporary a sustained consumption rise would not have been warranted. It was 1978 before OPEC had eliminated the annual increase in its cumulative surplus, following the first oil shock. As a result the transfer raised the average world saving rate, and contributed to a sustained fall in demand for the OECD. Two further sources of demand contraction can be identified. The rise in the price of oil pulled up other energy prices, thereby transferring income from consumers to energy producers (governments and private corporations) who it emerged had a lower propensity to spend on investment goods than the income losers, and finally there were tertiary spending adjustments by those groups gaining income within the OECD as a result of the spillovers from the first two effects. The net result was a substantial demand contraction in both oil shocks.

Much of the net real income fall subsequently emerged as unemployment in some of the OECD countries, and there seemed to be two basic factors at work here: real wage resistance and the subsequent investment contraction. As the price of crude oil rose for producers in the OECD, their unit costs increased and demand for other productive inputs, labour and capital, was diminished. To maintain employment, therefore, the oil price rise signalled the need for a fall in the real wage costs faced by producers as well as lower costs of capital. However, it is clear that the OECD economies differed in the extent to which real wages could fall, especially in the short run. To the extent that unions can speedily index wage rates to the general level of prices and obtain frequent (e.g. annual or semi-annual) contract negotiations then real wage rates tend to be sticky downwards. It is widely believed that whereas real wages can be lowered by inflation in the USA where two-year wage contracts are usual, in Europe and Japan real wage rates are likely to be much more inflexible. This real wage resistance will then give rise to unemployment when the economy goes through a supply side shock.

In fact the first two oil shocks differ in the degree of real wage resistance observed. Figure 2.5 examines the index of real labour costs relative to real national income in the USA, Japan and the major 7 OECD economies as a group. Since real national income drops in the oil shock aftermath, we would expect countries with flexible real wages to show a proportionate fall in labour costs. In these economies the index in question should show little change two years after each shock. This is clearly the case for the USA in both 1975 and 1981, and for Japan and the other major economies in 1981. However, in 1975, Japan and the group as a whole show, by the large recorded rise in the index, that real wage costs stayed up when real national income fell.

This real wage resistance, an important cause of unemployment in the mid-1970s, perhaps less so in the 1980s, also had the effect of squeezing out the share of profits in national income after the first oil shock. This, in turn, had the effect of sharply reducing investment spending, and so deepening the recession.

The first oil shock presented policy makers with the unfamiliar but unwelcome phenomenon of simultaneous accelerating inflation and rising unemployment. The resulting policy stances indicated considerable

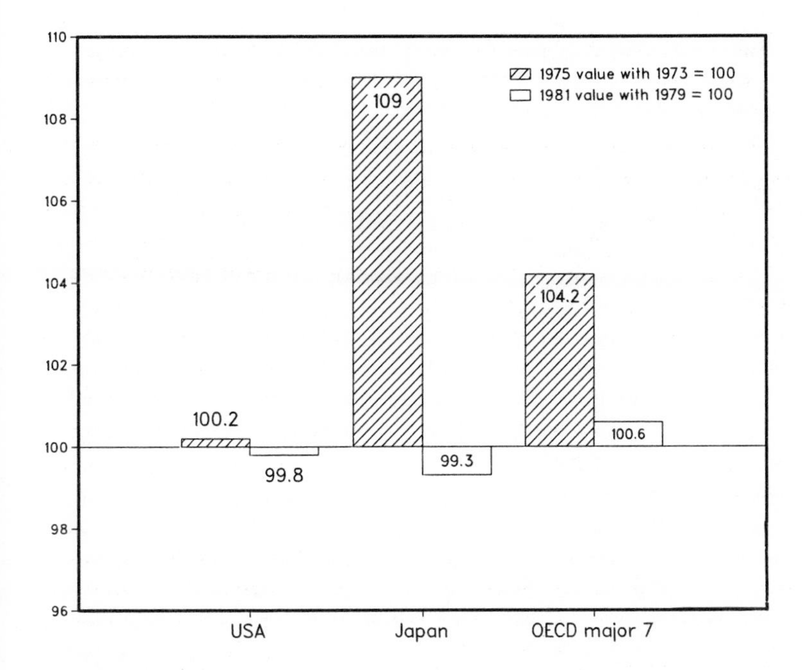

Figure 2.5 Real wage costs relative to national income
Source: OECD

uncertainty about the nature of the problem and how it should be tackled – an uncertainty which was less apparent after the second oil shock when the OECD governments were much more ready to ignore one half of the policy dilemma, unemployment, while tackling the second.

The different policy stances need to be considered in some detail, though, in part, the immediate policy situations associated with the oil shocks were predetermined by earlier policy and the automatic impact on monetary and fiscal aggregates of higher prices and rising unemployment levels.

Tables 2.1 and 2.2[5] set out brief histories of the policy stance following each of the major oil price changes in the major 7 OECD economies.

Before examining these tables it is necessary to know how fiscal and monetary policy actions can be measured.

In table 2.1 the indicator of the stance of monetary policy is taken to be the behaviour of short-term interest rates. The nominal short-term

Table 2.1 Monetary policy responses to oil shocks

Period	*Behaviour of short-term nominal and real interest rates in the major 7 OECD economies*
1973–5	Just prior to oil shock all 7 had been tightening monetary policy in the synchronized boom and the USA, Japan and West Germany all had nominal rates in excess of 11 per cent. Subsequently, in the USA nominal rates fell by more than half from 1973–7 as policy relaxed, although real rates rose again after 1975. For Japan, nominal rates more than doubled from 1973 to 1975, but real rates had dropped to −10 per cent by 1975. For West Germany, France and Italy both real and nominal rates fell over 1973–5. The UK's nominal rate stayed up, but its real rate plummeted from 2 to −10 per cent over 1973–5.
1979–81	Nominal rates in the USA generally rose but with large fluctuations. The real rate at first dropped sharply, then showed a very sharp rise. For Japan, the nominal rate rose sharply while the real rate continued its steady rise, begun in 1977. In West Germany, France, Italy and the UK there were sharp rises in nominal rates (the UK had started its rise in 1976–7 following the near collapse of sterling), and real rates also rose as inflation stabilized.
1982–5	There were generally positive and rising real rates through to 1985 as the USA budget deficit pulled up world interest rates and drew in money balances from outside the USA.

interest rate is what firms, banks and finance houses regard as the cost of borrowing money for the next few months. As oil prices push up the general level of prices this form of contingency demand for money will increase in the OECD economies, and if the stock of money is unchanged, short-term interest rates will rise to ration off the excess demand for short-term bank balances. Rising nominal short-term interest rates are, therefore, one way of showing that the government (or Central Bank) is not prepared to expand money and credit just because prices have risen.

On the other hand, the Central Bank may decide on an *accommodating* monetary policy. It therefore permits the money supply to rise in line with the higher prices so that there is no excess real demand for money. In this way interest rates are prevented from rising.

Table 2.2 Fiscal policy responses to oil shocks

Period	*Behaviour of government budget balances in the major 7 OECD economies; – = deficit; + = surplus*
post-1973	*Actual** balances moved into deficit from initial balance. *Structural** balances moved first to surplus in 1974 and then to deficit (approximately –1 per cent of GNP) by mid-1975. *IASBBs** outside USA began in surplus in 1973 and then moved steadily into deficit (about –2 per cent of potential GDP by 1977), while the USA reduced its surplus.
post-1979	*Actual* balances were about –2 per cent of nominal GNP by mid-1979 and moved further into deficit as shock developed. *Structural* balances of the group moved from deficit to balance by mid-1980 and then towards surplus. *IASBB* for the USA moved from surplus (about +4 per cent of GDP) to balance over the 1980–3 period, while the rest of the group moved from approximate balance in 1979 to a surplus of about +1.5 per cent of GDP by 1982. For the group the average *change* in IASBB over 1979–82 was +1.5 per cent of GDP.
post-1982	The USA continued to move into deficit on all three bases, and its *IASBB* was about –1 per cent of GDP by 1983; this was more than large enough to offset the approximately balanced position of the other 6. However, the latter was net of a substantial shift toward deficit in the UK and Canada. The net position of all 7 was an IASBB deficit from 1982–4.

* See text for definitions.

Higher interest rates can normally be expected to choke off investment spending, so adding to any supply side recession, but in this context it is necessary to look at real short-term interest rates; i.e. nominal rates less the rate of inflation. This is so because higher inflation adds to the money values of investment cash flows and if there is no simultaneous rise in the money cost of borrowing the real cost of a project will have fallen. If higher inflation therefore elicits a Central Bank response that keeps nominal interest rates constant, then real rates will fall, perhaps even becoming negative, and investment will be encouraged. Real rates may fall, therefore, when the Central Bank fails to anticipate the economy's actual rate of inflation.

Monetary policy had been restrictive prior to the oil shock to cope with the synchronized boom so that nominal rates were at historically very high levels in the USA, Japan and West Germany. In all of the OECD major 7 except Japan and the UK, nominal interest rates fell over the period 1973–5 as governments adopted more than accommodating monetary policy stances. Japan continued with its tight money policy to combat inflation while the UK continued to use the interest rate to manage its foreign exchange position. However, when real rates are examined, all of the major 7 allowed these to fall sharply.

The fundamental stance of monetary policy following the first oil shock was therefore to permit an investment and consumption expansion through lower real interest rates along with some accommodating falls in nominal rates, in order to offset the contraction in spending and therefore employment initiated by the real income transfer to OPEC.

Turning to fiscal policy the obvious indicator is the central government's budget balance, the difference between government spending on consumption and investment goods and in transfer payments on the one hand and its tax receipts on the other. However, three ways of measuring the budget balance must be distinguished.

The *actual* balance includes those automatic stabilizers of the business cycle that are triggered by the peaks and troughs in economic activity. As unemployment rises and incomes grow more slowly, more is paid out in unemployment benefits and social security payments. At the same time income related tax receipts must fall. The actual balances in all the OECD major 7 must therefore shift towards deficit (even if not into deficit) in any recession however caused, and this clearly did happen in both the first and second oil shocks. To eliminate this effect, the *structural* balance may be calculated. This measures what the net expenditure less receipts position would be if the economy were on its long-term growth and employment trend path, so that the structural balance is sometimes called the full employment budget balance.

However, inflation also affects the budget balance because it causes nominal interest rates to be raised as the money and financial markets seek to maintain a stable real rate of interest. Since nominal interest payments on the national debt form part of government expenditure, inflation therefore pushes both the actual and structural balances toward a deficit position. However, the other side of the coin is that inflation also erodes the real value of government debt denominated in money terms thereby transferring real resources from the holders of such debt (e.g. banks, insurance companies and private individuals), to the government itself. This is the so-called 'inflation tax' which may be said to reduce the real underlying tendency toward deficit. This can be allowed for by replacing, in the calculation of the structural balance, the actual interest payments by their counterpart in real terms. The resulting *Inflation Adjusted Structural Budget Balance* (IASBB) may be the best real indicator of the central government's fiscal policy position, as distinct from its current book keeping position.

Table 2.2 indicates that after the 1973 oil shock, the overall structural balance position of the major 7 moved toward deficit after an initial delay, while the IASBB having begun in surplus in order to control the pre-1973 boom moved toward deficit over the succeeding three years. The OECD measured the major 7's discretionary shift of fiscal policy towards deficit over 1973–5 to be nearly 2 per cent of GNP. This fiscal expansion even went so far as to include a lowering of the real price of oil products and gasoline as the specific taxes on these products were not raised in line with the general inflation. The fiscal policy signal after the first oil price rise was clearly that the supply shock and the failure of OPEC to spend its real income gains were reasons for expanding aggregate demand. In this respect Japan, and to a lesser extent, the USA were exceptions to the general rule of fiscal expansion.

Summarizing, the monetary policy stance for the first shock was to stabilize nominal interest rates, allowing real interest rates to fall sharply, and the fiscal policy stance was towards deficit on an IASBB basis. The analysis in tables 2.1 and 2.2 adopts the position that governments and central banks are conscious of the real outcomes of any policy changes denominated in money terms; in other words that falling real interest rates and deficit IASBB's were deliberately adopted. Nevertheless, it may be true that, given their relatively short-lived experience of rapid inflation in 1973, these real monetary and fiscal stances were not the conscious choice of governments, but only the consequence of their nominal policy decisions. Whatever the case, the outcome was the same.

Although the main purpose of expansionary fiscal and accommodating monetary policy was to resist the recessionary impulse of the terms of trade shift, particularly in those countries, like the UK, which could see a ready supply of import substitutes about to be commercialized in the North Sea, nevertheless, it had serious risks.

Given the degree of real wage resistance in the face of the supply shock, expansionary policy could do little to raise employment prospects through real wage decreases. All expansionary policy seems simply to have been translated into accelerating inflation. In turn, this general inflation itself weakened the real effect of higher energy price signals with the result that the required long-term substitution responses in production and consumption (i.e. conservation) were further delayed.

The general failure of this policy stance to do anything except add to inflation meant that in the second oil shock, policy was quite clearly different. In part it might be said that this was because the European economies in particular had elected, by 1979, governments more clearly from the right-wing political parties, and it might even be facilely argued that the failure of policy in 1974–5 set the scene for this political swing.

Whatever the case, both fiscal and monetary policy was not directed towards alleviating recession and unemployment in 1979–80. Indeed, in some cases the supply side shock to employment might even have been welcomed as a further backup to the general commitment to demand contracting anti-inflation policy that had become the overriding objective in most industrial countries.

As table 2.1 indicates monetary policy was tightened and the major 7's Central Banks permitted nominal interest rates to rise and as inflation became contained this permitted rising real rates of interest as well, choking off consumption and investment.

Table 2.2 indicates that the real income losses of the second oil shock automatically pushed actual budget balances into deficit. However, the major 7 governments maintained a contractionary fiscal stance by pushing their structural balance toward surplus. At this time the USA began to expand its budgetary policy on an IASBB measure but the other six moved towards surplus on the IASBB basis, none more so than the UK which had for 1979–82 a net movement of its IASBB towards surplus of 6 per cent of GDP (Llewellyn 1983).

The environment in which the second oil shock occurred is therefore notable for several differences from earlier periods. In the industrial economies there was an elevation of the economic objective of price stability over the importance attached previously to other objectives such as growth, and full employment. At the same time, governments started to pay more attention to monetarist views about policy making. These emphasized two principal ideas: first, that economic policy can only affect inflation through demand management, and, secondly, the most effective form of anti-inflationary demand management was a restrictive monetary policy. As a consequence real economic growth was constrained and interest rates rose generally throughout the OECD economies. The OECD calculated that 'the cumulated real GDP shortfall between 1979 and 1982 was due as much to policy tightening as to the oil price rise itself.'

This pushed the industrial countries into severe recession. The oil price rises reduced the productivity of the industrial economies' productive factors leading to a noticeable productivity slowdown. Since this was itself inflationary the policy response was towards further monetary contraction. The resulting unemployment did reverse the decline in productivity but was clearly a consequence of restrictive demand management. The recession and unemployment however prevented a re-emergence of the real wage resistance which characterized the first shock, as figure 2.5 indicates. All in all therefore the restrictive monetary stance largely paved the way for prolonged recession and unemployment on the one hand, but successful containment of, and adjustment to, the inflationary consequences of the oil price rises.

Because of this squeeze on real wages, profit shares held up much better in the second shock with the result that investment spending was able to withstand some of the restrictions induced by the high real interest rates.

The fiscal and monetary policy stances were designed to reinforce each other, with some governments, notably that of the UK, arguing that restraint on public expenditure was necessary to facilitate control of the money supply. The OECD referred to the trend towards deficit of actual budget balances as one of the problems of macroeconomic imbalance rather than a natural consequence of recession.

Despite the actual budget balance's automatic move to fiscal expansion induced by recession, the major 7 governments became alarmed at the swelling deficits and Llewellyn (1983) suggests three reasons. Firstly, it was feared that the structural elements, i.e. the committed non-cycle related welfare payments and transfers, were growing permanently larger. Secondly, it was argued that as government spending rose and hence so did its borrowing requirement this increased the pressure on capital markets to lend to government at the expense of lending to private industry. This pressure showed up as higher interest rates with the result that private investment was 'crowded out'. Lastly it was argued that a growing GDP share for the public sector was a drag on the growth prospects for the economy as a whole. Although the latter argument only has merit in so far as the public sector adopts investment projects which the private sector would reject as uneconomic and most OECD economies have public sector investment appraisal rules which guard against this, nevertheless governments, in the general monetarist climate, did seek to contain public sector growth.

The maintenance of a trend in fiscal policy towards underlying budgetary surplus on an IASBB basis in a period of recession clearly indicates that Keynesian policies of demand management to offset the business cycle were out of favour at this time. Although the initiating cause of the downturn was a supply side shock in the form of the oil price rise, by concentrating on the inflation impulse and ignoring the stagnation effect, government policy in the major OECD economies severely deepened the recession. The contrast with 1973–4 is clear in that the earlier shock had been the occasion for governments largely to ignore the inflationary consequences of their policies to offset the unemployment consequences of that supply shock.

The OECD has also drawn attention to the problem of current account deficits in the balance of payments, and figure 2.4 indicates how current accounts moved for three principal groups: the oil exporters, the industrial countries and the nodc's. Broadly, OPEC accumulated surpluses in the aftermath of each shock, while the industrial countries moved into deficit, especially prolonged after 1979. The nodc's began in deficits and these worsened considerably over the period 1973–83.

The accumulated current account surpluses of OPEC prolonged the recession in output after each shock, but their counterpart was a capital inflow account surplus to deficit nations with a corresponding capital account deficit to OPEC.

In the initial analysis of a terms of trade shift it became clear that one adjustment response of the damaged country was to reduce imports and increase production of import substitutes. Hence there is no reason why a terms of trade shift should produce more than a temporary shock to the current account. Other things being equal, the initial deficit should be quickly eliminated.

But other things were not equal after the oil shocks, and to understand this it is necessary to reflect on what a current account deficit signifies about trends in the economy.

In the fixed exchange rate era a prolonged current account deficit was usually indicative of a country's inability to pay for its imports by competitive export production, and deficits were usually financed by official borrowing until eventually the exchange rate was devalued, along with some demand contraction, in order to re-establish competitiveness. However, this view of current account deficits as only signalling uncompetitiveness is often misplaced in a world of floating exchange rates and highly mobile capital.

At least two other interpretations of the current account are possible.[6] Under a floating exchange rate system, the exchange rate appreciates or depreciates to bring the overall balance of payments into equilibrium. This balance can include a position of current account deficit and accommodating capital account surplus over a long period, as the country in question borrows on private capital markets rather than from official sources.

The first interpretation of a current account deficit is that it reflects an excess of spending on goods and services over the domestic production of goods and services. This spending may be for both consumption and investment and is commonly labelled absorption. The excess of absorption over income, reflected in the deficit, may be financed by borrowing for consumption and investment abroad through the capital account. Following the first oil shock, many oil importers were uncertain whether the oil price rise and the resulting loss of real income would be permanent or only a transitory shock. Those countries who believed it to be transitory may have felt that there was no reason therefore to reduce their long-run permanent consumption levels. The deficits on current account simply reflected their desire to finance a continuing level of consumption through temporary borrowing. This effect was reinforced by those countries which wanted to spread out the decrease in consumption levels which a permanent loss of real income would signal. This was undoubtedly important for those countries with the highest import bills, so that for a short time after the oil shocks there was evidence that the size of current account deficits was positively correlated with the degree of oil import dependence of the countries in question.

A second interpretation of the current account deficit focuses on the investment rather than consumption component of absorption. The current account deficit is then taken to reflect the excess of domestic investment over national saving (both public and private). The offsetting capital account inflow is clearly then a positive decision by world capital markets to invest in the country running the deficit. Far from being a sign of uncompetitiveness, the current account deficit is then a sign of optimistic investment opportunities.

Both of these interpretations clearly fit, for example, the UK economy after 1973–4. The policy stance was towards offsetting the drop in consumption and rise in unemployment signalled by the real income loss, and as oil prices rose the investment opportunity afforded by North Sea oil improved as well. There were, therefore, reasons for both temporary deficit financing of consumption and for investment inflows. The investment inflow interpretation was particularly relevant in the case of Norway. Sachs (1981) reports that, after 1973, Norway's ratio of investment to GNP was nearly 38 per cent, but since the savings ratio remained stable, this was financed by a current account deficit that had approached 13 per cent of GNP by 1977, without any discernible rise in the cost of capital to Norway or depreciation of its exchange rate. The deficit disappeared as the investment projects came to fruition.

The opposite case applies to Japan and West Germany after 1973. In these countries the policy related deep recession in conjunction with the profit squeeze consequent on real wage resistance appears largely to have reduced the ratio of investment to GNP, resulting in current account surpluses. This was reversed in 1979 when profit shares held up, investment was maintained and Japanese and West German surpluses decreased.

On this view, the argument that deficits reflect permanent overspending or uncompetitiveness has been criticized as mercantilist.

Nevertheless, the OECD concern about the deficit position of the oil importers was a serious one. Essentially the total OECD deficit was a necessary counterpart of the OPEC surplus which resulted from OPEC not spending its real income gains. This represented an overall increase in the world saving rate and hence permitted a fall in world interest rates with consequent stimulus to investment reflected in the oil importers' deficits.

However, there remained the problem of the distribution or allocation of those deficits amongst the oil-importing group as a whole. Clearly OPEC wished those countries offering the best investment returns (e.g. from energy saving investments) to run the largest deficits, but following each shock it was those countries most dependent on oil imports which needed to run the largest deficits. The two groups of countries were not necessarily the same.

This revealed the problem of petrodollar recycling, which for many governments in the 1970s was the real embodiment of the oil crisis. The problem of how to recycle OPEC's surpluses from OPEC's desired investment targets to those countries most in need threatened to undermine the international financial system.[7] The world capital markets had to reconcile lenders who wanted very short-term, risk-free, high-return homes for their funds with borrowers who needed long-term and, in the case of the nodc's, low-cost loans in relatively risky circumstances.

There is some feeling that the official international monetary agencies coped less well with this crisis than the private banking systems of the world. Nevertheless, it is certainly true that the sort of competitive devaluations to shift deficit adjustment on to other countries that characterized the 1930s and that were a real fear of the OECD and IMF were avoided.

In the event international lending and intermediation by the private banking systems expanded enormously in the 1970s to cope with the recycling problem. The OPEC surpluses were rechannelled through the industrial countries, and, in the view of the IMF, they were in this way able to replace the industrial countries' own savings as the main source of funds for the nodc's. The international banking system used the industrialized countries simply as a route for the shift of OPEC surpluses into nodc's current account deficit.

In the process, it appears through the 1970s that while the deficit countries were able to avoid running down their reserves, the oil exporters were able to increase their holdings of short-term, risk-free reserve assets, chiefly dollars. This net reserve creation seems largely to have occurred through the expansion of the Eurocurrency markets, and in particular the willingness of the USA to run both current and capital account deficits as it relaxed its own controls on capital exports and acted as lender of last resort to the world.

By the time of the second oil shock this system had matured and proved capable of handling the recycling problem. Nevertheless, the position of the nodc's had deteriorated substantially, and their overall response to the oil shocks is worth separate consideration.

The non-oil developing countries

The history of the oil shocks has been different for the nodc's in several significant ways.

It has already been observed that these countries began rapid industrialization after 1950 and began to take over from many of the industrialized countries as exporters of, especially labour intensive, manufactured goods. This industrialization effort and the drive to raise living standards partly explains the long-term commitment of the nodc's to expansionary monetary and fiscal policies. As a result, inflation in this group exceeded that in the rest of the world in all periods after 1950, as figure 2.2 shows. Indeed it can be seen that the slowdown in the world inflation rate after 1980 was due entirely to restrictive policies in the industrial countries, while the inflation rates in the nodc's as a group continued to accelerate.

At the same time, as figure 2.4 indicates, this group's current account deficit did not adjust after the oil shocks in the way observed in the industrial countries. The deficit continued to accumulate up to 1981, and since then, despite some improvement, it has remained at about the level observed at the time of the second oil shock.

This accumulated deficit with its capital inflow counterpart led eventually to what has become known as the third world debt crisis. Several interpretations, both optimistic and pessimistic, are possible.

One view is that the nodc's deficit simply reflects the shift in investment opportunities away from the industrial countries to the developing economies. If this is so, then the deficits will attract a capital inflow to projects earning a real rate of return in excess of the world interest rate. There is then no problem of long-term repayment just as there was no long-term problem for Norway in its own deficit financing.

On the other hand, if the deficit represents a sustained excess of absorption for consumption, and does not attract a capital inflow directed at improving real growth, then a long-term debt crisis clearly exists. One aspect of this problem is that in the aftermath of the recycling crisis the burden of providing loans and therefore monitoring credit worthiness of individual countries shifted from the official aid agencies whose national government sponsors were cutting public expenditure, to the private capital markets. These, it could be argued, were collectively inexperienced at evaluating the optimum lending programmes for the nodc's. Some economists have argued that the investment shift from industrialized to nodc economies has itself been directed by the lower rate of return expectations in industrial economies as demand management policy has been restrictive. This is in contrast to the generally expansionary stance of policy in the nodc's which boosted investment return expectations.

It is noteworthy that one of the most problematic debtors has been Mexico where large capital inflows were used – along the lines of the Norway model – to develop indigenous oil and gas resources just at the time when the third oil shock, with falling oil prices, came along.

The nodc debt crisis has importance for Europe for several reasons. Firstly, European banks are among the largest creditors of the nodc borrowers; secondly, individual European member states have particularly close links with their former colonies in the nodc group, and, thirdly, the EEC itself through the Lomé Convention and through such broad informal initiatives as the Brandt Commission has a particular interest in the growth prospects and aid requirements of the nodc's.

Clearly the nodc's were even more seriously undermined in terms of growth and inflation by the familiar 1973–4 combination of oil shock and offsetting monetary expansion than the industrialized countries. The combination was repeated after 1979 at a time when the industrialized countries were restraining demand growth. In this respect, some economists have characterized the 'North–South' hemispheric division in the following terms.[8] The North, largely the industrialized countries, represents a broadly demand-constrained economic grouping; that is to say policy has been directed at restrictive demand management allowing excess supply to develop in the form of unemployment. The South, on the other hand, largely the nodc's, represents a supply-constrained economic grouping where policy has been directed at maintaining aggregate demand in the face of supply shortages so that inflation accelerates. This suggests that there is a coordination problem at the level of the world economy, exacerbated by any protectionist tendencies in the North that restrict the industrialized countries' ability to provide markets for the output of the South.

In the event, it appears that the ability of the nodc's to attract further capital inflows was diminishing by 1983, and as a group they were forced to contract their current account deficits by their own restriction of imports and aggregate demand, choosing deflation rather than default as a response to the debt crisis.

Weaker oil prices and the world recovery

By 1983–4 the third oil shock, a reversal of the first two, was clearly imminent.

Firstly, the long-term economic responses to the terms of trade shift were bearing fruit. Increased indigenous energy production and severely curtailed energy demand amongst the industrialized countries largely weakened the demand for OPEC oil, and the resulting excess production capacity in OPEC weakened the cartel's own market power. Secondly, the aggregate demand restraint in the OECD, despite its unemployment consequences, had produced short-term support for the long-term reduction in demand for OPEC oil – enough to permit some of the OECD economies, including the UK, to shift the stance of fiscal policy measured by IASBB back towards expansion (see table 2.2). The combination of weaker oil prices and mild fiscal relaxation began to pull up the growth rates of the industrialized countries by 1983–4. There was no accompanying surge in oil and energy demand, however, because (as chapter 4 shows in detail) there had been a considerable conservation success in the industrialized countries (including those of the EEC) resulting in a lower energy and oil intensity of real national income. That is to say that energy and oil consumption growth no longer bore the same proportional relationship to real income growth that had been observed prior to 1973.

At first the nominal oil price moved down slowly as OPEC found itself with a smaller and smaller share of falling world oil consumption. In real terms the oil price held up through 1984–5 as OECD export prices fell even faster. However, by the end of 1985 demand for OPEC oil was 5–6 million barrels per day below capacity and dissension in the group meant the so-called low absorbers were no longer willing to restrict production for the sake of the others. The details of OPEC developments are set out in chapter 3, but, in summary, the cartel broke up (though perhaps only temporarily) when, in 1986, its members separately began to aim for increased production, price cutting, and a re-establishment of the pre-1980 market shares. From that point on, as figure 1.1 indicates, nominal and real oil prices dropped steadily. By early 1986, the real price of oil denominated in a basket of currencies was less than 20 per cent above the level it had been at the end of 1973.[9] The whole of the second oil shock and most of the first, measured in terms of income shifts, were wiped out.

Analysis of this third oil shock, in part, is the converse of the first two, and requires examination of the effects on four groups of countries: the industrial oil importers, the industrialized energy producers, the oil exporters and the nodc's. As before three aspects stood out:

(a) Would there be an asymmetric spending effect if the losers (OPEC chiefly) contracted their spending faster than the gainers (the industrialized oil importers) expanded theirs?
(b) What would happen to inflation?

(c) How would governments respond in adjusting monetary and fiscal policy?

The primary effect was to lower inflation rates and the general level of prices, especially in the industrialized oil importing countries in the EEC, such as West Germany, France and Italy and in Japan. Coupled with these lower prices would be a growth in economic activity unless the oil importers were particularly slow in raising their expenditure after receiving the real income transfer represented by the improvement in their terms of trade.

Part of this boost to economic activity depended on the stance of macroeconomic policy. If monetary policy was accommodating so that governments did not allow the lower prices to raise the real value of the money stock in their economies (i.e. they had a contractionary monetary policy), the boost to economic activity would last for only a short period. Any attempt to offset the effect of lower prices in implicitly raising budget deficits would also cut short the stimulus to economic activity. Thus, while inflation performance would clearly improve, there were many qualifications about the extent to which governments would allow the fall in oil prices to raise economic activity and employment. Over the longer run, the lower oil prices could certainly be expected to bring about some increase in the overall production potential of the oil importing economies.

On the basis of OECD calculations[10] it can be suggested that a 40 per cent fall in oil prices, such as occurred in 1985–6, could lower OECD inflation by about 5 per cent. If monetary and fiscal actions permitted the resulting implicit expansion, OECD economic activity would be nearly 2 per cent higher within three years, assuming import demand by OPEC's low absorbers held up. On the other hand, reactive monetary and fiscal contraction to hold interest rates stable combined with a cut in OPEC import demand might even cause a decrease of 0.5 per cent in OECD economic activity.

Within the OECD, a second group consists of the energy producers whose supply effort was stimulated by the first two oil shocks: the USA, the UK, Canada, Norway and so on. The same initial effects of lower prices, higher spending and higher production potential would apply to them. However, particularly in the case of the UK, these would be offset by depreciations in their exchange rates (especially against the Deutschmark and Yen) since falling oil prices would lower their oil revenues and require them to find replacement sources of export earnings. In the case of the UK, the British Treasury estimated that exchange rate depreciation (with consequent inflation) would be the dominant effect. The policy dilemma for these countries would be whether to permit the inflationary consequences of depreciation or to further raise interest rates in the pursuit of price stability. Another strong policy temptation was the use of oil product taxes to replace the loss of government receipts from oil revenues.

For the nodc's such as Brazil and India the reduced pressure on inflation and the potential for increased worldwide economic activity and trade was a most welcome addition to the direct saving on oil import bills. If the implicit monetary expansion amongst the industrialized countries was allowed to lower interest rates then this would even help their debt servicing problems.

The principal losers were the oil-exporting developing countries, such as Mexico and Venezuela. Debt servicing problems and income losses would be the most severe consequences, and might even threaten the collapse of those western banks which had lent heavily to these countries. On the other hand, the boost to the prosperity of the west could allow it to take a more lenient view of the debt repayment problems of the third world.

In general, therefore, a discrete shift downwards in industrialized country prices was the general outcome of the 1986 OPEC break up. Whether this led to overall greater economic activity in the short run depended on the willingness of governments to permit the lower prices implicitly to ease their monetary and fiscal stances, and permit interest rates to fall. Over the longer run, any aggregate demand constraint would be offset by the increased supply and production potential following lower energy prices in oil-consuming countries.

As in each of the other oil shocks, a significant redistribution of income and wealth had occurred and whether or not this changed the level of world economic activity depended critically on the asymmetry of responses between gainers and losers and their governments.

Finally, there is the question of whether the (temporary) collapse of OPEC would permit a growth-led return to pre-1973 levels of oil consumption. Two obstacles to this existed: the degree of conservation and substitution already set in motion by the first two oil shocks, and the changed perception in oil-consuming countries of the likelihood of cartel action in oil prices.

3 Europe and the world oil market

The oil shocks brought considerable attention to bear on the world's energy demands and supplies, and the role of Europe and Europe's fuel markets emerged as critical. The rise in oil prices, which by 1986 remained at historically high levels despite some real erosion over 1982–5, pulled up other energy prices, so that conservation and substitution processes have been observed in all fuel markets.

Several issues related to Europe's role in world energy markets are important. First, there is the set of conditions which permitted OPEC to raise oil prices and that subsequently kept oil prices high. These conditions include both the factors determining OPEC's behaviour and the factors that caused Europe to be vulnerable to oil shocks.

Secondly, there are the political economy responses to the oil shocks: by this is meant the emergence of the oil importers as a strategic group. The group embarked on several initiatives: alliances, confrontation or accommodation with OPEC, crisis management, formation of the IEA and so on. Disparate views of these developments need to be accounted for, including the EEC's perception of its role.

Thirdly, there is a need to understand the nature of the long-run conservation and supply responses to the oil price rises. These two aspects are the subjects of chapters 4 and 5 , and this chapter concentrates on the first two issues: cartel power in the world oil market, and the consuming countries' responses.

Developments in the world oil market

In the light of the oil price experiences of the 1970s, it seems natural to think of the world oil market as one from which competitive pressures are absent, and always have been absent.

Many commentators are however agreed that a few broadly defined epochs may be used to categorize oil price developments over the post-war period, and these include periods when competitive pressures clearly determined prices.

Approximately, the oil market epochs may be said to span the post-war period, 1947–58, the cheap oil era, 1959–69, the OPEC decade 1970–80, and the conservation response 1981–6,[1] but these are only very broad characteristics since, for example, OPEC itself came into being as early as 1960.

THE POST-WAR PERIOD 1947–58

The post-war period is usually regarded as one of relatively tight oil company control of the world market, and it is from this era that the pervasive notion of a company cartel dominated by the Seven Sisters, (Exxon, BP, Shell, Chevron, Texaco, Gulf and Mobil) arises. In fact, the major oil companies who both produced crude oil and refined petroleum products in their own vertically integrated operations had operated as an effective cartel since the 1930s.

During this first era, world crude oil prices were geared to the Texas Gulf prices of indigenous USA production through the basing point system. The c.i.f. price of delivered crude oil in Europe, for example, was set at the f.o.b. Texas Gulf price plus a schedule of shipment costs previously agreed on by the cartel members. This applied to all oil whatever its source or the company delivering it. Indeed the companies integrated their operations to the extent of delivering each other's oil to minimize transport costs while maintaining the Texas Gulf f.o.b. price basis.

Since by 1948 the USA was already starting to import oil, while production was simultaneously expanding in the Middle East, the artificial basis of the price system was open to challenge. Nevertheless, it survived for most of the 1950s, since by having a relatively high 'posted price' based on Texas Gulf rather than the much lower Persian Gulf costs, the oil companies as refiners could apparently justify high prices for petroleum products, and at the same time permit the extensive profit sharing revenues to their Middle East host governments which all felt helped the political stability of the region.

As Persian Gulf output expanded the companies shipped this oil both to Europe and the USA. Had there been a competitive market structure, the new lower Middle East f.o.b. price would have taken over as the marginal cost of crude oil, and prices in each market would have fallen to the sum of shipment costs and the Middle East f.o.b.

price. However, the company cartel appears to have been able to avoid lowering all its prices, and by discriminating between its European and American customers, continued to base European prices on the Texas Gulf f.o.b. price plus shipment costs, while offering the lower prices only on the American shipments. On several occasions European governments and the Economic Cooperation Administration that was financing oil imports through the Marshall Fund were able to discover that their c.i.f. prices implied a higher f.o.b. price (or 'netback') on shipments to Europe from the Middle East, than was implied by American c.i.f. prices, and this led to an antagonism towards the oil companies which partly found expression in the objective of protecting indigenous coal producers in Europe. It is arguable that this form of protectionism subsequently committed many governments to a continuing interest in high oil prices.

The company cartel with stable prices geared to the Texas Gulf could not survive the onslaught of a variety of economic developments in the 1950s. The most important of these was the huge expansion in Middle East production capacity as the uniquely rich oil deposits of Saudi Arabia, Kuwait, Iran and other Middle Eastern countries were developed under the well-organized exploration activities of the major oil companies.

This Middle Eastern oil was very much cheaper than any of the American reserves. In a definitive study, Adelman (1972) calculated, in the mid-1960s, the *maximum economic finding cost* of Middle East oil to be approximately 10–20 cents per barrel, about 5–10 per cent of the realized f.o.b. price that had been set on world markets by the oil companies in the 1950s. The available reserves of this crude oil were so large that at prevailing prices there was clearly a chronic excess supply of oil for world markets.

A second factor diminishing the power of the cartel was the influx of new entrants into oil production and refining. The 'independents' from the USA, and the national oil companies of Europe undermined the ability of the cartel members to go on stabilizing world oil prices. The classic cartel problem – large profits from concentrated operations leading to competitive new entrants unwilling to participate in the cartel – emerged to erode the majors' grip on world markets. The new entrants, which included the government owned integrated oil companies of France, Italy and Belgium, were able to obtain considerable concessionary operations in the Middle East countries. The additional crude oil supplies they were able to add were diverted largely to European markets because in 1959 the US government, under pressure to protect indigenous high-cost independent oil producers, instigated oil import quotas. At the same time, the USSR, with its growing oil production, began a price-cutting policy in Europe, encouraging the widespread erosion of the old company cartel pricing structure.

In 1960, the oil companies accepted the emergence of the competitive pressures and lowered the posted prices on which they based payments to their host governments and their own c.i.f. crude oil prices in Europe. This diminution in their profit share just when their revenues were expanding was a deeply felt injustice to the host governments. After years of prompting from Venezuela they formed in 1960 the Organization of Petroleum Exporting Countries (OPEC) with the aim of re-establishing the high oil prices of the 1950s and gaining control of oil production in their own countries.

THE CHEAP OIL ERA 1959–69

The era of cheap oil which competitive market pressures brought about saw a substantial switch from coal to oil consumption in Europe as figure 3.1 indicates.

The system of payments by the oil companies to the host governments in the oil-producing regions had begun in 1950 as a fixed payment per barrel. As production expanded in the 1950s this was supplemented by an income tax on profits earned and companies and governments settled on a '50–50' split. In the USA such payments were tax deductible for USA income tax purposes partly to smooth the transfer of subsidies to host governments in order to maintain the political stability of the region.

The taxes themselves required prices at which oil sales were valued, and these 'posted prices' had initially been set in the late 1950s at the actual selling prices at which Middle East crude oil was transacted.

As the excess supply of oil grew and competition flourished these posted prices though still used for tax purposes became more and more discounted in actual sales. Finally, it was the reduction in these posted prices in 1959 which signalled the companies' attempt to diminish their tax payments.

During the cheap oil era of the 1960s it was OPEC's aim to re-establish their revenues by stabilizing posted prices, altering the basis of taxation and in the long run gaining control of production and pricing itself. The first step was to cease treating the fixed per barrel

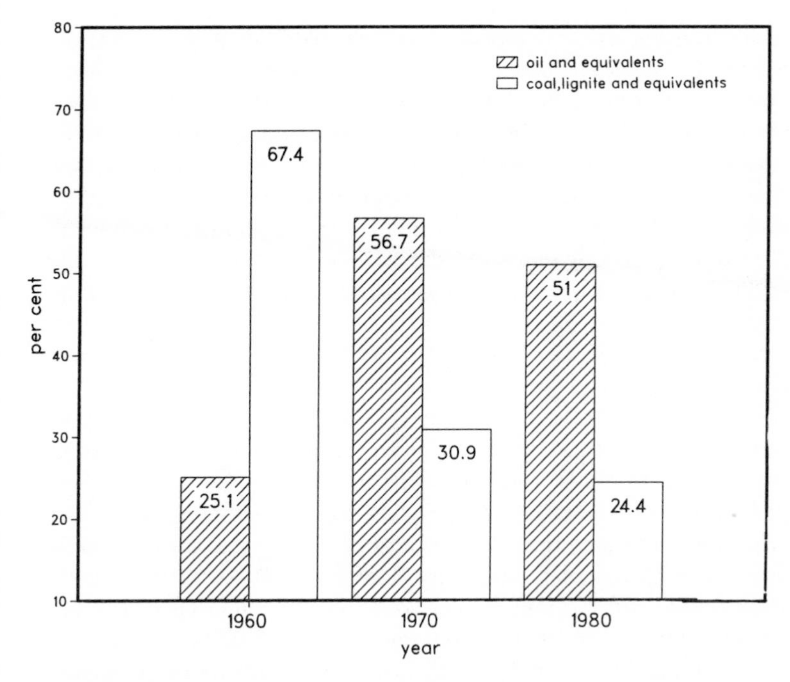

Figure 3.1 Shares of oil and coal in EEC EUR9 primary energy
Source: Eurostat

royalty as part of the income tax, and insisting instead on it being a cost to be subtracted before the 50 per cent tax rate was levied. In this way the floor to oil prices became the marginal cost plus tax, and the effective cartel power was transferred to OPEC since its member governments, notably led by Libya, were prepared by the end of the 1960s to impose their tax rate proposals under threat of nationalization.

It would appear therefore that by 1969–70 OPEC was able to stabilize its earnings despite cheap oil prices and could be ready to consider the next step towards participation leading to eventual total control of its own oil production capacity. The Saudi Arabian oil minister Sheikh Yamani hinted at this in 1968, but it is interesting to note the comments of Adelman writing in 1972. He argued that division of Middle East oil profits need not be a zero sum game so that whatever the oil companies lost, OPEC gained. He based this on the view that the companies and OPEC together could form an efficient cartel with the companies acting as OPEC's tax collectors, and the public record of tax receipts acting as a check on any member's attempts to cheat the cartel. He argued that if the oil companies were eventually reduced simply to being buyers of oil from governments it would be more difficult for OPEC to detect cheating or price discounting in a weak market. A dozen years later the truth of this became apparent. It took less than this to see the folly of ignoring another of Adelman's arguments: that 'would be to the immediate and ultimate benefit of all – a stockpiling program to render Europe, Japan and the rest of the OECD nations safe against a concerted shutdown' (Adelman 1972, p. 215).

THE OPEC DECADE 1970–80

By 1970 world demand for crude oil had escalated strongly in the wake of the synchronized boom in OECD countries. At the time the relatively long experience of stable fuel prices lulled buyers and sellers alike into imagining that oil and energy demand would always grow in tandem with economic prosperity. However, despite the knowledge that OPEC still felt unfairly excluded from the large rates of return that were being earned on the extremely low cost Middle East oil, even experienced observers did not feel that OPEC could do much to push up prices in view of the abundance of low-cost supplies.

Nevertheless, a start was made when, exploiting a temporary rise in oil prices on the Rotterdam spot market that arose because of a tanker shortage in 1970, Libya raised posted prices and the income tax rate, and was then followed by other OPEC members. The resulting negotiations, the Teheran-Tripoli Agreements of 1971, established OPEC's ability to set posted prices and tax rates itself by threatening to deny oil to non-complying companies and set the scene for moves toward greater OPEC participation in company operations.

Despite the general agreement that the dominant characteristic of the world oil market for many years has been the potential *excess supply* of oil at the prices ruling in the 1970s, that decade saw the success of OPEC in being able to institute short-term panics about *supply shortages*, followed by massive price rises.

Between 1970 and 1973 several factors came together which enabled OPEC to wield considerable monopoly power. The boom in the OECD countries continued and the nodc's supplemented the increasing oil demand with their own expansionary policies together with improved terms of trade from a commodity price boom.

On the supply side, OPEC was able to make use of the large number of independent oil-producing and refining companies which had entered Middle East production in the 1960s. Libya was able to take the lead in this by threatening to restrict the independents' (including the European national oil companies) access to further oil concessions, so threatening their market shares in the oil-importing countries. The continued threat of loss of access enabled OPEC to lead up the posted prices of the independents and serve notice that a new era in oil prices was to emerge.

In 1973 the Arab–Israeli war caused even the apparently moderate Arab states in OPEC to contemplate using oil as a means of influencing US and European foreign policy towards their causes, and embargo arrangements were announced for supplies of Middle East oil to the USA and the Netherlands. In fact the embargo would not have been a possibility without the close relationship between OPEC and the major oil companies, and the *de facto* avoidance of the embargo was also only possible because the integrated operations of the companies permitted them to share the misery equally amongst the consuming nations.

In this environment of temporary production cutbacks, bidding up of spot oil prices on the Rotterdam market by independent producers temporarily worried about access to their concessions, and the extreme inelasticity of European oil demand in the very short run, OPEC was able to bring about the first of the oil shocks.

Essentially two types of response to the oil price rise were possible. On the one hand many argued that this was only one more short-term demand panic in an otherwise long-term history of falling oil prices, and that OPEC would go the way of every other cartel in internal bickering and eventual break up. On the other hand there were those who felt that the oil price rise signalled the start of an energy shortage and that both permanent conservation and investment in high-cost alternatives would characterize the future.

There were significant spillovers into other fuel markets and energy prices rose over the remainder of the decade, though not always in real terms. However from 1974 to 1978 oil prices remained stable, and the world's preoccupation was with the macroeconomic responses to the oil shock.

In 1978–9 short-term factors were again able to permit OPEC to push up prices abruptly. The fall of the Shah of Iran and a temporary loss of Iranian oil production coincided with a large rundown of oil stocks in Europe and rising spot prices. It took very little in the form of production cutbacks to stimulate another panic about oil shortage and although the International Energy Agency's (IEA's) emergency allocation scheme was now available, it was not brought into operation. The OPEC cartel was able to push through the second oil shock.

THE CONSERVATION RESPONSE 1981-6

Energy conservation took off substantially as the cumulative effects of two price shocks and the policy-induced recession caused oil demand in particular to collapse. The glut of oil at prevailing prices became apparent, and for the first time OPEC was faced with maintaining prices through production cutbacks. This job was made more difficult because the Middle East members had completed their moves towards nationalization and none of the major or independent oil companies could be used to allocate production quotas instead of explicit compromises within OPEC. Indeed, by acting simply as buyers of crude oil the oil companies were now countervailing competitors rather than collaborators in OPEC's control of world oil markets.

As the price of oil was eroded by OPEC's failure to maintain complete solidarity, the third oil shock, the price relapse, became evident. Between 1980 and 1985 spot oil prices fell by about 30 per cent, and spot market transactions by this time accounted for nearly half of all international oil transactions, but for European consumers the lower oil prices were delayed by the strength of the American dollar, the currency in which oil prices are denominated. By 1985 the fall in the value of the dollar was allowing substantially falling oil prices throughout the world, and persuading OPEC to consider setting prices in terms of a currency basket. At the same time, OPEC had to give serious thought to internal auditing and supervision of members' production and prices in order to avoid price discounting. It even contemplated arrangements based on the Texas Railroad Commission which the American oil companies had used as an effective cartel instrument in the 1930s, and in October 1984 specific production quotas were set for OPEC members, to be audited by the group, in order to detect cheating and price discounting.

Competitive pressures in the world oil market

There is no evidence that oil supply is a natural monopoly – one of those industries where it is cheapest to have only one large supplier –

and there have been periods of intense competition in oil supply, notably, in recent times, the 1960s, and perhaps again in the 1980s. Nevertheless the first two oil shocks arose through the large-scale use of monopoly power to escalate prices above levels at which there was already a long-term excess supply. It is clear therefore that relatively complex forces may be at work in the oil market.

Accepting that the long-term trend in oil prices is essentially competitive but punctuated by bursts, sometimes quite prolonged, of monopoly power to raise prices, it is necessary to ask what can cause monopolistic groups to arise in oil supply.

Economics suggests several factors which make the cartelization of oil supplies a likely possibility at repeated intervals.[2] First, the short-run price elasticity of demand is low so that consumers cannot reduce oil consumption quickly or substantially when price rises, and the price rises themselves are extremely profitable for producers able to bring them about. This inelasticity chiefly arises because oil has few substitutes, and these often require considerable time lags before their supply can be increased. In the long run the demand may be much more elastic, but over a period of weeks or months very substantial rises in spot prices might be needed to ration the consequences of a temporary disruption in supply. If such temporary disruptions can be used to stimulate fears of oil shortage over longer periods, price rises can be sustained. In order to carry out these price effects producers need to collude together as a monopolistic group.

A second factor may arise from the geographical or political proximity of a group of suppliers who find that natural obstacles to collusion are absent. Whether they will act as a cartel (as the oil companies in the 1930s or OPEC in the 1970s) depends on their perception of the profits to be gained from concerted action. Some estimates suggest that OPEC as a group increased its permanent wealth by about 100 per cent as a result of the 1973 price rises. Such high absolute gains to cartelization were sufficient for even the least benefited members to avoid cheating in the OPEC decade.

A third factor is whether, after price has risen and consumption starts to slump, over the long run, the cartel is prepared to maintain unity, by allocating production cutbacks and quotas. In the case of OPEC, so strong was the demand for oil even midway through the 1970s as the OECD continued to support aggregate economic activity, that the cartel could simply turn the allocation problem over to their tax collectors, the oil companies themselves. With their globally integrated operations spanning different Middle East producers, the companies probably came relatively close to the most profitable (for OPEC) allocation system anyway by concentrating production on lowest marginal cost areas. The absolute level of additional wealth going to all members was large enough for them not to have to bother with their own production quotas. Hence absence of specific OPEC quota mechanisms is not evidence that OPEC was not a monopolist in the 1970s, as some have argued.

High cartel profits however are the signal for new entrants to the industry and both the 1960s and the 1980s saw a proliferation of competitors operating on the fringe of the company and OPEC cartels. Such competitive fringes supplying substitutes for the cartel's product are the most usual cause of break up. The outcome is a cyclical process whereby the potential for large profits encourages cartelization and price rises, but the actuality of large profits encourages competitive entry and price cuts.

A final factor permitting cartelization may be failure of consumers to act together to break the cartel. After 1973 the oil companies stayed on as OPEC tax collectors rather than consumers and prior to the oil price rise, the OECD nations had failed to agree on the stockpiling policy which could have frustrated OPEC's repeated short-term shortages of supply. Consuming countries individually cannot gain so greatly from their own stockpiling policy since an individual stockpile cannot usually prevent world prices rising, and can therefore offer only a temporary alleviation of price rises or embargo.

Eventually the majority of OECD nations formed the International Energy Agency (IEA) and along with the EEC, most of whose members were in the IEA, the former seemed to offer a countervailing power to OPEC. Of the possible policy responses to OPEC, three seemed to stand out:

(a) stockpiling and emergency allocation
(b) taxes on oil imports or consumption
(c) subsidies to and protection of indigenous energy production.

OPEC as a cartel

Three views of OPEC as a cartel have generally been on offer over the last dozen years or so – setting aside OPEC's own statements of its long-term aims.[3] The first of these sees OPEC purely as a wealth maximizing joint monopolist conscious that its main source of income is subject to considerable volatility as markets respond, and governments in the consuming countries adapt their fiscal and monetary policy to price shocks.

A second view, and one that was extremely popular in the 1970s, holds that the OPEC members have different target revenues to achieve. Some members, the high absorbers, are aiming for profit maximization of low oil reserves to raise the living standards of their large populations. Others, the low absorbers, have only limited 'needs' for current oil revenues since their populations are low and they take a very long-term view of the future – or, in the language of economics, use a low discount rate for future consumption. These countries might even cut back production as prices rise, reversing the usual view of producers' market responses.

A third view has it that OPEC's behaviour can be seen entirely in terms of geopolitical objectives and that oil price rises are merely incidental instruments of individual national foreign policies. Two difficulties with this argument are that it is, firstly, very difficult to disprove by evidence of market behaviour since it can include any *ex post* rationalization of economic actions and, secondly, there is no reason to believe that it adds to or detracts from the purely economic explanations offered above. In other words, even if OPEC's objectives were purely political it is difficult to see how sacrificing additions to wealth would enable it to be in a better position to achieve political objectives. Indeed, one of the major casualties of the OPEC decade was the idea that special political relationships between countries could prevent any of the consequences of exploitation of monopoly power.

The first two views are therefore of primary interest, and associated with the second must be mentioned the idea that OPEC was not really a causal factor in raising oil prices, merely the transmitter of the market signal that energy scarcity was becoming serious. In its most popular form, as frequently emphasized by OPEC representatives, the OECD countries were to blame for oil price rises because of their profligate disregard of energy conservation.

The argument that OPEC consisted of two separate groups of low and high absorber countries seemed in the 1970s to add weight to the view that oil prices could only continue to rise in real terms. As figure 3.2 shows, OPEC countries contain the bulk of proved oil reserves in the world including centrally planned economies (CPE), and within OPEC, 65 per cent of its reserves are owned by the so-called low absorbers: Saudi Arabia, Kuwait, Qatar, Libya and United Arab

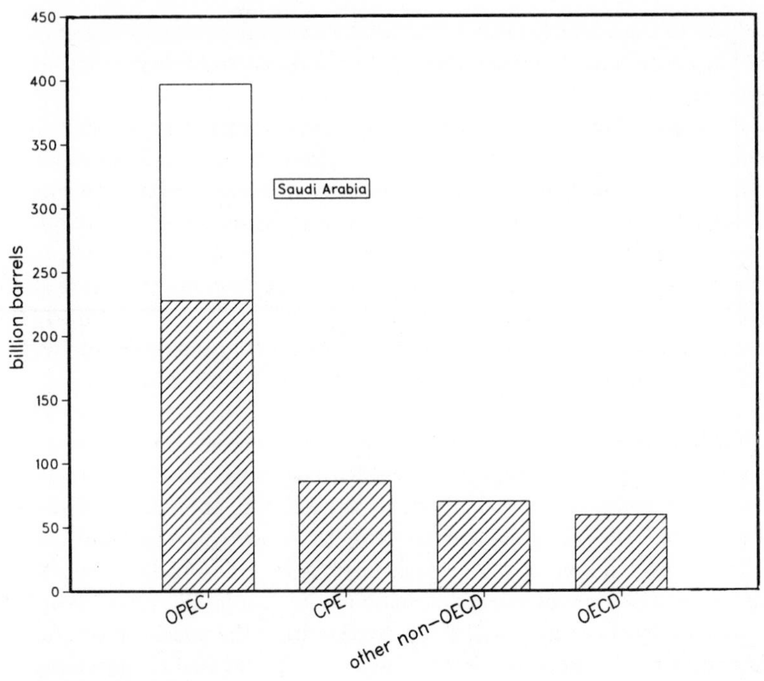

Figure 3.2 Proved oil reserves, 1984
Source: Exxon Corporation, IEA

Emirates. This group also possessed more than half of OPEC's installed production capacity (figure 3.3). In turn, the low absorbers were dominated by Saudi Arabia with 38 per cent of OPEC reserves (25 per cent of total world reserves) and 32 per cent of its production capacity.

With this leverage over reserves and production capacity the low absorbers could clearly put long-term world crude oil supply under threat if they were unwilling to maintain output over the long run. For this group it was often suggested that oil in the ground was their best investment. The basis of this argument was the economic theory of exhaustible resources.

When an owner of a finite resource that will be exhausted by continued production is contemplating depletion policy, the comparison of present and future prices net of extraction costs is critical. The usual expression for price net of extraction cost is *royalty*, and as exhaustion approaches and the supply of the resource falls compared with the demand, the royalty which can be charged will rise. The resource owner,

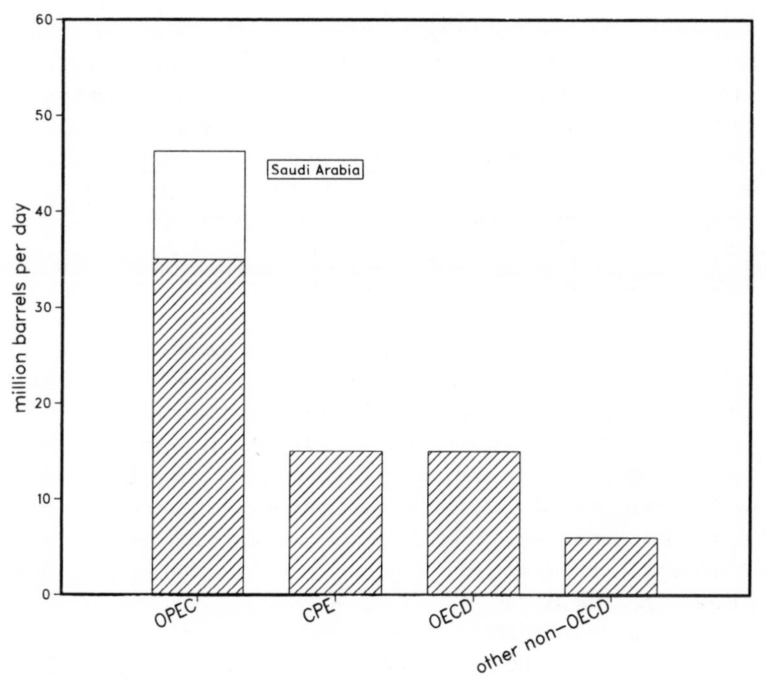

Figure 3.3 Crude oil production capacity, 1981
Source: IEA

in contemplating the current output level, knows therefore that future royalty will exceed present royalty, but this fact alone is not enough to cause a restriction in current production. The future royalty carries with it an additional cost – the cost of waiting for its receipt, so that production decisions then require the comparison of the current royalty with the *discounted present value* of the future royalty.[4] The lower the discounted present value of future royalties, the more likely the resource owner is to deplete the resource immediately in order to turn the resource into some other earning asset.

Two factors determine the discounted present value of future royalties: the price that oil is expected to fetch in the future compared with the present, and the discount rate used by the oil producers. The proponents of the idea that OPEC had a decisive core of low-absorber nations argued that these countries had a very low discount rate for delayed future consumption, so that they hardly valued current receipts more highly than those for which they would have to wait a long time. This low discount rate, it was occasionally argued, arose because they actively wanted to delay western-style industrialization. In this model, the OPEC core had no incentive to increase current production unlike the fringe of high-absorber nations with large populations seeking immediate consumption gains and discounting the future heavily. The result would be a non-cooperative oligopoly with the strength of the core keeping oil prices high even in a weak market. OPEC's own public face was not in conflict with this view since it presented the apparently attractive picture of a group conserving a scarce resource for future generations.

This view of OPEC as a non-cooperative oligopoly of high spenders and low discount rate conservers has not survived the 1970s well, and a simpler model in which OPEC is a collusive wealth maximizing cartel with a high discount rate has several arguments in its favour. Most striking amongst these is that, as figure 2.4 indicated, the oil exporters used up their post-shock current account surpluses within 2–4 years by increased consumption. Such increased consumption levels, once established are not easily reduced, especially by governments carrying out prestige projects, and by 1981 the oil exporters were maintaining their consumption by current account deficits.

A second factor is the volatility of oil revenue flows, especially after 1979 when OECD governments adopted restrictive macroeconomic policy. Adelman has argued that what then determines the discount rate is not so much the resource owner's own preference for present over future consumption but the additional risk from holding oil assets compared to relatively risk-free paper asset alternatives, with the result that all oil exporters' discount rates are increased.

A third argument in favour of the cartel model of OPEC is that the theory of exhaustible resource economics may simply not be applicable to crude oil at this time. Since the excess supply period of the 1960s, reserves have continued to be discovered, consumption has stabilized and market prices provide a greater incentive to explore. The reserves to production ratio in the world oil market has remained in the 25–35 year range for the last quarter century, and by 1983 was at a 20-year high. In these circumstances the effect of escalating exhaustion royalty on market price may be negligible, so that the enormous gap between market price and marginal extraction cost (a factor of about 100 in 1983) can only be the result of collusive monopoly power.

However, the profits of cartelization always encourage a competitive fringe, of which three types can be identified in the world oil market.

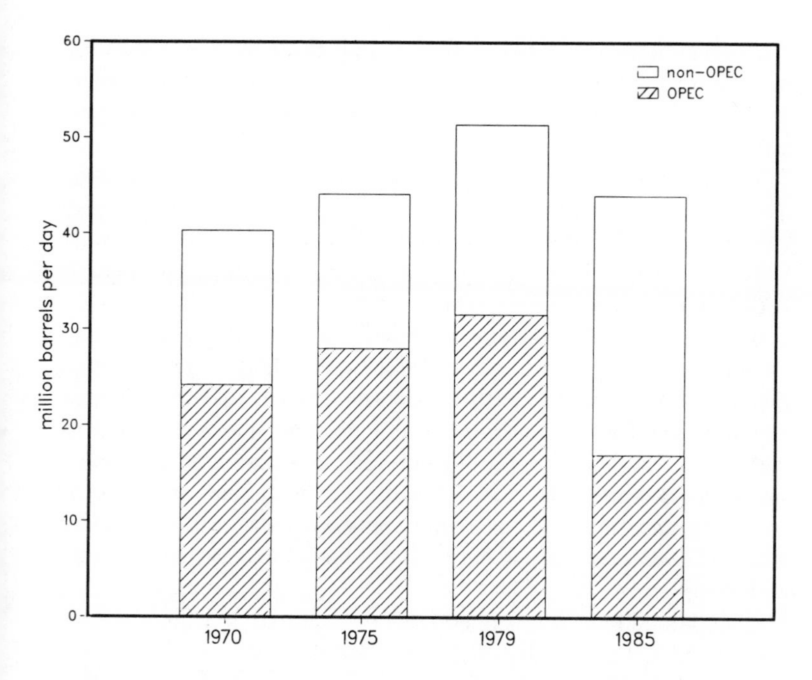

Figure 3.4 Crude oil production in the world outside Communist areas
Source: IEA, The Economist

The most obvious fringe, and the one blamed by OPEC for the price erosion of 1984–6 consists of the OECD oil producers including the UK and Norway. While figure 3.2 indicates that their reserves are in fact insufficient to make them more than price takers in the world market, nevertheless figure 3.4 is able to show how, along with the rest of the non-OPEC fringe, they moved in to fill the gap in the declining crude oil market when OPEC tried to limit its production to about 16 million barrels per day (mbd) in 1984–5.

The second fringe may arise within OPEC if the pressure of falling sales and prices on consumption levels does cause the cartel to degenerate into a non-cooperative oligopoly. In 1984–5 there was considerable difficulty within OPEC about discovering the degree of cheating on price discounts in the weak state of the world market.

The third source of competition comes from 'backstop' technologies of energy supply made more economic by high oil prices; nuclear power and, in a sense, the initially underestimated degree of energy conservation observed in the 1980s are the major examples.

Cartel strategy in the face of fringe competition is classically a form of limit pricing, i.e. to let market prices drop for a period in order to see off the fringe and then to attempt to raise prices again, if the cartel can hold together in the interim.

To cause the most rapid depletion of non-OPEC reserves might require a prolonged period of low prices, and presumes a cartel strategy based on a low discount rate and a great deal of patience. On the other hand this strategy also undermines the backstop-conservation fringe. It is the strategy advocated in public by OPEC.

In 1986, however, most observers saw the price erosion as a symptom of the cartel's break up or a threat by the OPEC core against its own partners. A policy of cutting prices temporarily to discipline those members who had lost sight of the large present value of cartel profits is consistent both with high discount rates amongst core and fringe, and with the fact that Saudi Arabia had lost patience with its role as the swing producer in a weak market – compared with its capacity of 11 mbd and 1981 peak output of 10 mbd, its 1985 production was only 2–3 mbd.

In deciding, in its December 1985 meeting, to aim for increased market share, OPEC adopted the netback pricing system used successfully in the 1950s by the company cartel to disguise the relationship between prices and costs from its consumers. The role of netback pricing in 1986 clearly fulfilled the requirement of disguising the extent of price undercutting within OPEC itself.

By 1986, when real oil prices were only about 20 per cent higher than at the start of 1974 (see figure 1.1) the outlook was very uncertain. A successful sharp lesson from the OPEC core achieved quickly, along with a resurgence of world demand, could have pushed oil prices up again fairly rapidly. On the other hand, the break up of the cartel under the onslaught of three types of fringe competition could mean that limit pricing turned into one of the oil industry's prolonged periods of competitive low prices. A major difference between 1986 and 1960 was that oil importers' perceptions of the risks of oil dependence were significantly different.

European energy policy before 1973

In Europe, the substitution in the 1960s of oil for coal and other energy sources produced a variety of government responses, many of which reflected official antagonism towards the company cartel.

The basis of the antagonism took several forms. First, there was an underlying view that the company cartel administered an excessively profitable pricing structure. There had been a foundation for this in the habit in the 1950s of tying delivered crude prices to Texas Gulf prices plus transport costs. However, by 1960 the cartel-fringe effect of new entrants competing prices down as demand expanded had effectively removed the company cartel as the dominating force in price setting. It is arguable that had the US government not banned imports this factor might have proved an even stronger incentive to erode prices.

A second source of antagonism was the fear that the oil companies wielded excessive power in allocating European energy resources, and as a result there were strong pressures for control to be wrested from them by the government of the consuming countries. In 1964, for example, the EEC formulated the objective of achieving community ownership of crude production for Europe, along the lines of the French model.

France had a long-established policy of national control of oil distribution with a requirement to use 90 per cent from French refineries. Further upstream, France had its 'own' supplies of crude from Algeria and the basis of *le petrole franc* was to ensure through French-owned refineries, an outlet for French-owned crude. By the late 1960s, Compagnie Français des Pétroles (CFP) was as active as the majors in searching for further oil reserves, and in particular a form of protectionism for Algerian crude grew up. France was prepared to pay high prices (along with specific aid packages) to obtain secure supplies of Algerian crude to French refineries. The high crude prices in turn were used to justify high ex-refinery prices for petroleum products.

A combination of French government oil companies (ERAP) extended this policy by searching for additional expensive concessions in other Middle East countries. France's oil policy was therefore aimed at establishing a secure supply of high-priced crude for French refineries through special political and economic relationships with particular producing countries. This policy set the tone for a generalized interest in high oil prices in Europe according to some economists. Indeed, Adelman reports that in 1967 Italy's ENI, and ERAP along with some German oil companies, sought EEC aid and protection from the competition of the international oil companies.

The interest in high oil prices was reflected in nearly all the other European countries by attempts to protect the indigenous coal industry, but the attempts appear only to have been partially successful even in West Germany and the UK.

In West Germany, a national oil company, Deminex, was set up to be first a buyer then an explorer for crude oil, but it proved unable to challenge the position of the major oil companies in the penetration of the home market, and an essentially competitive market in oil product distribution developed in West Germany.

The coal industry had been largely rationalized after the Second World War and the old coal cartels broken up with the aid of the ECSC. By the 1960s a private company Ruhrkohle AG had been adopted by government, with large shareholdings by electricity utilities, as a single coal supplier. Oil and petroleum products were taxed, coal production subsidized, and coal imports restricted.

Essentially the same policies – oil taxation, coal subsidy and restriction of coal imports – were adopted by the UK (and indeed Japan) and Adelman has commented that coal protectionism has been the oldest and most important aspect of energy policy in Europe.

The particular characteristics of European deep-mined coal supply seem to have contributed to the objective of protecting it. The industry is labour intensive with production concentrated in particular areas which become dependent on coal field employment as the main source of income. The close-knit communities which develop militate against labour mobility in and out of the industry, and there are strong pressures to isolate wage rates, the main component of costs, from market pressures on the demand for the commodity. Delaying the inevitable rundown of coal supply for social reasons therefore became an overriding objective in those countries which had an indigenous coal base.

In addition to protecting national oil companies, and protecting coal against oil, a third source of protectionism in the UK was the desire to support the growing nuclear power industry. In this respect, the UK had followed an even more nationalistic line than most other European users. While France and West Germany were prepared to manufacture US designed pressurized water reactors (pwr's) under licence in national nuclear industries, the UK had pursued its own gas-cooled technology which had proved hideously expensive. In the UK especially, as well as in Europe as a whole, the desire to protect high-cost nuclear power as a bulwark against international oil company power contributed to an ethos of high oil prices being acceptable.

In Italy, the development of an internationally competitive national oil company, ENI, had been one of the main factors in the challenge of the independent producers to the majors' company cartel in the 1960s. Again the incentive was for the national oil companies of Europe to bid up concessions to ensure access to Middle East oil, not to lower prices, but to block the company cartel.

In summary, the emphasis in energy policy in Europe prior to 1973 had been to erode the role of the international oil companies by establishing national oil companies geared to high oil prices, and by protecting high-cost indigenous oil substitutes. The proposed objectives were cheap but secure energy. In the event neither was achieved. The aim of security of supply led to a preference for high-cost sources of fuel, while the cheapest way to ensure security, the stockpiling of low-cost Middle East oil, was never seriously evaluated on an EEC-wide scale until very late in the day. By attempting to restrict demand for oil despite its market penetration in the 1960s, Europe may have frustrated (as the USA did) the loss of cartel control at the centre. As it happened cartel control passed relatively smoothly from companies to OPEC.

This review has emphasized national energy policies, and these can be compared with their collective reflection in the Community as outlined in the description of the historical background in chapter 1.

Oil shocks and the European response after 1973

Although macroeconomic policy dominated the initial response to the oil shocks of the 1970s, attention rapidly had to be given to formulating oil and fuel market responses as well.

Two broad strategies emerged: on the one hand, some countries, most notably France, favoured the maintenance of special political relationships, as had happened in Algeria after French withdrawal, to ensure secure or favoured oil supplies. On the other hand, the consumer countries sought to break the OPEC cartel as quickly as possible, despite their failure to prevent its emergence before 1973.

Maull (1980) in his large-scale study *Europe and World Energy* set out the arguments for the first approach. Arguing that political processes are the chief explanations of world oil price trends he felt that the importance of oil for the consuming nations far outweighed the significance of oil revenues for OPEC. A continued disruption of world oil could lead to worldwide economic and political changes affecting the OECD, OPEC and the nodc's alike. In this event, the rewards from cooperation plus the costs of non-cooperation outweighed the benefits of market determined outcomes for all participants.

The opposing view is best characterized by the comment of Adelman (1984) that no political consensus could ever be expected between the supply and demand sides of a market, and that the failure of such strategies as US President Carter's cultivation of Saudi Arabia to prevent oil price escalation after 1979 is evidence for this.

The conciliatory and the confrontational strategies ran alongside in the aftermath of the first oil shock. During 1974–5 considerable emphasis was given to the idea of consumer–producer consensus. Under the leadership of France, the principal oil-consuming nations, OPEC and the nodc's convened in Paris in 1976 a Conference on International Economic Cooperation (CIEC). Despite considerable support amongst European oil importers, this strategy did not attract favour in the USA or the UK where indigenous energy supplies were still large. The CIEC in fact failed to confront the oil market issue, and instead concentrated on the North–South disparity in world prosperity. OPEC allied itself with the nodc's in seeking a new international economic order for primary producing countries, a theme subsequently taken up by the *Brandt Commission*, a private study group on the North–South issue.

In the USA, the emphasis was much more on building what Maull referred to as 'an economic NATO' – the International Energy Agency (IEA), principally the brainchild of US Presidential Adviser on National Security, Henry Kissinger. The enduring achievement of the IEA was its emergency oil allocation scheme, designed as an improvement on the 1973–4 arrangements.

The 1973–4 oil embargo on the USA, the Netherlands and Denmark had been left to the oil companies to implement, on behalf of OPEC, and to allocate for the OECD. The EC Commission subsequently carried out a detailed review of oil company procedures during the embargo. It found that member states had put their own needs first, with each trying to pressurize its national oil companies to maintain its own supplies irrespective of the rest of the Community.

In addition, member states introduced maximum refinery product prices which had the effect of squeezing out many of the smaller independent petroleum companies. In general, however, the Commission did not find that allocations had been significantly different during 1973–4 from a year earlier, and concluded that the oil companies, who on their own initiative decided to share the misery equally, had been in as much difficulty as their customers. Nevertheless, the Commission noted the complexity of ex-refinery price systems and their markup over crude oil costs, and adopted as a long-term aim the establishment of *transparency* in Community fuel pricing.

The IEA was set up in 1974–5, and consisted of most OECD countries, the most notable exception being France. Its emergency allocation scheme, still in place in 1986, required member states to hold 90 days' worth of oil imports in stock; in 1984 this averaged about 17 per cent of annual consumption for EEC members. The objective is to avoid selective member states being embargo targets by sharing the disruption around – however the most oil-dependent members suffer most.

If one or more members is embargoed so that for the IEA as a group there is a 'shortfall' of at least 7 per cent of normal consumption the scheme may come into operation unless voted down by the IEA Governing Board. A 'shortfall' is the difference between actual supplies and 90 per cent of normal consumption. If activated, the scheme then requires members to implement the 10 per cent consumption cutback implicit in the formula, and each member's supply right is equal to the difference between its restricted consumption and its permitted share of stock drawdown. This supply right comes from non-embargoed imports, its own production, if any, plus an allocation of the production of any net exporters in the group (e.g. the UK and Norway), the allocation being determined by the member's share of normal IEA total imports.

Yager[5] reports that the Iranian revolution of 1979 brought about a 5 per cent shortfall to the IEA members as a group, but no triggering of the scheme was activated, with the situation simply being handled by a request to members to consider consumption cutbacks of about 5 per cent. The oil companies simply reallocated other oil to those members who had previously imported large amounts from Iran.

The stock holding position remained ambiguous for most of the 1970s since different countries held their stocks in different ways, largely in commercial oil company inventories, and the 'bottom of the tank' was difficult to measure because a minimum stock level is necessary to keep the distribution system functioning. The EEC members (aside from France) did however agree to subsuming the Community's stock holding policy (90 days of *consumption*) into the IEA scheme.

Two doubts remain about the IEA scheme. Many observers agree with Maull's view that there could be attrition of membership in any real emergency, especially if it was found that countries due to receive allocations had not been scrupulous about maintaining stocks. Moreover the question of how oil prices should be allowed to rise in an emergency has not seriously been addressed either by the IEA or the EEC. The 1973–4 episode showed a mixture of price control mechanisms, and while there would be pressure on governments to use non-price rationing in at least part of the market, allowing prices to rise in an embargo is itself the most efficient way of balancing supply and demand.

The notion of politicizing energy markets through a confrontational IEA did not survive beyond the 1976 US elections, and President Carter adopted a much more conciliatory attitude especially to Saudi Arabia (which in turn suggested that building up the US strategic Petroleum Reserve was unfriendly). Attention shifted in the IEA, in the economic summits of the OECD major 7, and in the EEC towards energy conservation.

In particular, a long-running topic of policy debate was the idea of taxing oil imports, and it is worth examining this idea in a little detail, especially since it was not unusual in the aftermath of the second oil shock to hear expressed the view that because the west had not conserved oil by taxing it, it had contributed to its own vulnerability.

The security of supply premium and the oil import tax

Consider how a tax on crude oil imports would work. The first effect is to reduce both the volume of imports and therefore the revenues passing to OPEC. Secondly, where there is already some indigenous production of oil the supply will expand as the price within the country imposing the tariff rises above the previous world price. Thus while a part of OPEC's revenues is destroyed a part is transferred by the higher domestic prices to indigenous producers (i.e. North Sea oil) in order to cover their higher production costs. Thirdly, resources are transferred from consumers in two ways. On the one hand the tax proceeds on the higher oil or oil products are paid to government which may return them in the form of lower taxes elsewhere or higher government spending (for example, on energy investment). On the other hand, the higher domestic prices transfer resources from consumers to indigenous producers which not only covers their costs but yields some profits or rents to these producers.

The net effects are: lower imports, lower OPEC revenues, higher payments to indigenous producers, and two sources of real income *loss* which will not appear in the national accounts: (a) the loss of real income to consumers who now have to pay more for each barrel of oil consumed than before, and (b) the loss to the economy as a whole arising from the substitution of high-cost indigenous production for relatively lower cost imports.

If the tax results in these real income losses, it is necessary to ask on what grounds it has been suggested.

Two fundamental arguments seem to underlie the frequent calls for oil import taxes, and the suggestion that the tax would reflect a social risk premium over and above the world price of oil.

The first argument is one consistently favoured by many US economists and is based on the hypothesis that US and European oil imports (which in 1983 accounted respectively for 18 and 37 per cent of world crude oil trade) are together large enough to bestow *monopsony* buying power. This means that if US and European consumption dropped, the effect on world demand might be large enough to force OPEC to lower its prices. The USA and Europe together or even separately might break up OPEC's monopoly by a countervailing exercise of monopsony power to restrict consumption and lower prices. In that event, the argument goes, the net real wealth of the USA and Europe might increase, though retaliation could be expected. This 'optimum tariff' case falls into the category of *economic* arguments for a tariff, i.e. arguments that the tax will increase the taxer's net wealth.

A second argument is based on the case that the probability of supply disruptions rises with the volume of a country's imports, and that such disruptions impose avoidable economic costs. Neither hypothesis is unchallengeable. There is no evidence that import volume increases the probability of disrupted supply (the 1973–4 embargo was not aimed at the most oil-dependent countries in any of the usual senses). Nevertheless, this second argument is the basis of a long-term preoccupation with *security of supply* as an objective of energy policy, one that has featured in policy objectives of the IEA, the EEC and their member states for many years.

It is clear that markets are used to dealing with commodities whose supply is uncertain or subject to random shocks. Both indigenous non-oil producers and consumers have views of the probable price that will be required to ration demand in an absence of imports, and combining this with the price ruling in normal times, they will form an *expectation* of the price that will rule over the long run averaging out embargoes and normal supply. Consumers will base their risky consumption levels on this expected price, as will indigenous producers deciding on their production effort. Both together have the incentive to establish stocks of the risky commodity, with the marginal cost of stock holding reflecting the additional security premium.

Economic analyses (e.g. Tolley and Wilman 1977) have suggested that in risky situations consumers will build stockpiles until the marginal cost of storage is equal to the difference between the price that would clear the home market in a supply disruption and the world price that rules normally. The security of supply premium is, therefore, this price difference multiplied by the probability of disruption. This, in turn, when added to the world price that would normally rule gives the expected long-run marginal cost of oil which risk neutral consumers and producers use to determine their demand and supply. Security of supply objectives therefore does not necessarily imply any policy is needed over and above private market responses.

This conclusion is incorrect when there are external costs of disruption not taken into account by private market decisions. Many have argued that the macroeconomic costs of supply disruption such as inflation and unemployment fill the bill in this context, and that an import reducing tax is needed to pre-empt such disruption induced costs. Others have argued that macroeconomic consequences are the result of failure to allow market forces to work as rapidly as possible and that an oil import tax is not the appropriate policy.

What is clear is that after 1973–4, a considerable change in the general view of the probability of disruption occurred and while private market precautions may have been inadequate prior to 1973, these were certainly increased subsequently.

Proposals for additional oil import taxes were particularly important in 1979–80 when the EEC Energy Commissioner, Guido Brunner, proposed a 1–2 per cent increase in the price of a barrel of imported oil with four objectives:

(a) to cut consumption
(b) to reduce the UK share of the Community budget
(c) to provide the EEC with a new 'own resource'
(d) to finance spending on new energy sources.

Objective (b) would transfer resources to the UK's oil producers, and

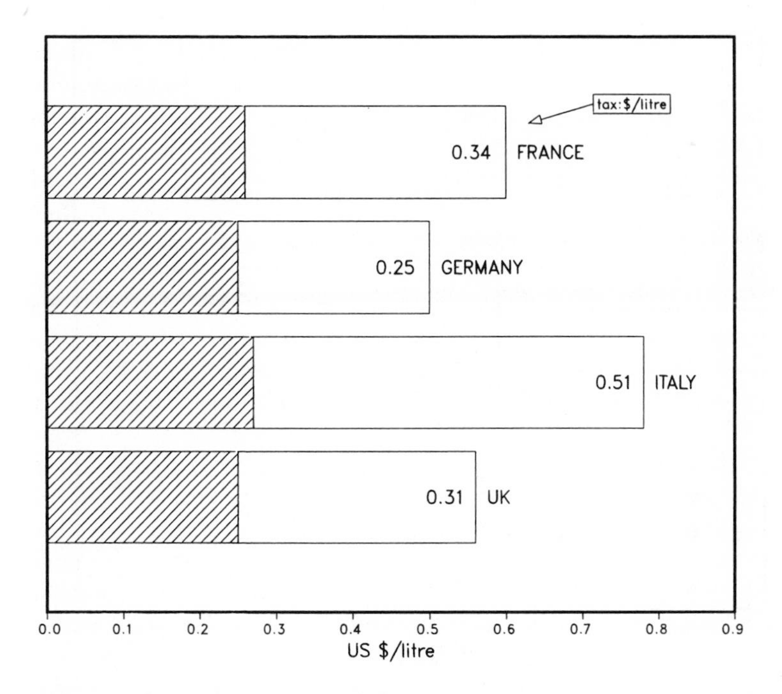

Figure 3.5 Prices and taxes of petrol in the EEC, 1984
Source: IEA

the UK government would have recouped the profits via a windfall profits tax. However, the proposal, the only serious contender for a supranational Common Energy Policy in the 1970s was not agreed upon.

In 1986, the oil import tax argument arose again in the face of falling oil prices particularly in the USA. There, the independent US oil companies who would largely benefit from it, argued for an oil import tax as a way of funding the US budget deficit from the revenues of OPEC, at a time when oil prices would otherwise fall in both nominal and real terms.

In fact, oil taxes always have been part of the European economy, and figures 3.5 and 3.6 indicate the existing 1984 petroleum product taxes in parts of the EEC.

What became clear by 1984–5 was that the oil price rises alone had been overestimated by OPEC, and that the conservation response and the indigenous supply expansion in European fuel markets turned out

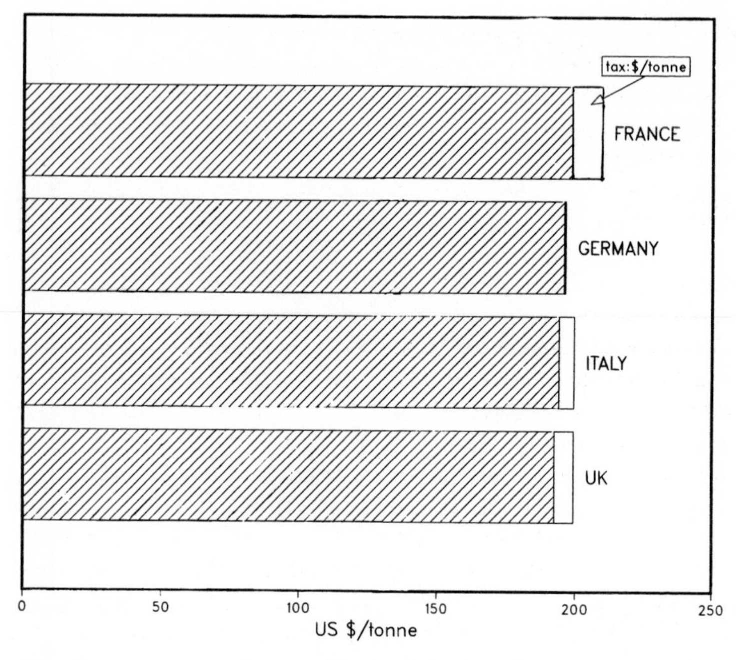

Figure 3.6 Prices and taxes of heavy fuel oil in the EEC, 1984
Source: IEA

to be much larger than expected. Certainly large enough to break up the cartel without the use of either confrontational or conciliatory relations or specific import taxes. These responses are the subject of the next two chapters.

4 Energy conservation and demand in the European Community

The oil shocks examined in detail in the last two chapters were primarily responsible for raising the European consciousness about energy. The oil price rises pulled other fuel prices up and long-run conservation and supply responses resulted. This chapter concentrates on the Community's demand response and conservation effort.

A major part of the Community's response has been the adoption of sets of energy policy guidelines.

EEC energy policy objectives

The first set of guidelines was called exactly that and appeared in 1968. They expressed a desire to 'guarantee long-term security of supplies under satisfactory economic conditions', but little of substance besides this all-embracing objective. The first oil shock coincided with the Commission's first forecasting exercises, and from this point on the objectives became much more specific. Three sets have been issued: 1975 for 1985, 1980 for 1990 and 1985 for 1995.[1]

The 1975 for 1985 objectives, *Towards a New Energy Policy Strategy for the Community*, made use of the Commission's forecasting exercises, and covered both demand and supply. The demand objectives were, as was fashionable at the time, couched in terms of improving the consumption performance that was forecast to occur if no policy changes were made: i.e. a comparison with a business-as-usual forecast. The drawback was that such business-as-usual forecasts are frequently in error by such horrendously large amounts, that the objectives become rather meaningless. The Community's performance was no better in this category than that of many of its member states.

The demand objectives were, first, to reduce by 10–15 per cent the 1973 forecast of what consumption would be in 1985 and, secondly, to achieve a share for electricity in final energy consumption of about 35 per cent by 1985. These were to be in the context of an overall share for imports of not more than 50 per cent, though an independent estimate by two Commission members (Brondel and Morton 1977) thought that 50 per cent dependence was over-optimistic given the trend of member states' policies. Implementation of the objectives was to be left to member states.

Table 4.1 indicates that the demand objectives were rather easily achieved mainly because the initial demand forecasts were so grossly in error, a characteristic of nearly all energy forecasts made in the pre-oil shock era.

On the supply side, it was hoped that indigenous energy supplies would achieve the specific target levels shown in table 4.2 (along with actual figures for 1983), although both Denmark and the Netherlands expressed reservations about the proposed expansion in nuclear capacity.

In general these supply projections varied from the mildly to the wildly optimistic, except for the huge overestimate of import requirements for natural gas and oil. These latter reflect the Commission's understandable inability to predict both the growth of North Sea oil (which does not count as an import) and the large contraction in energy demand compared with the forecast level.

The 1980 for 1990 set of objectives more sensibly were specified in terms of improvements on past actual performances, rather than on hypothetical business-as-usual forecasts.

In summary the 1980 for 1990 objectives were as follows:

1. to reduce oil dependence, i.e. oil's share of primary energy consumption to about 40 per cent (compared with the 1980 share of 52 per cent)
2. to ensure that 70–75 per cent of inputs to electricity generation were from nuclear power and solid fuels (compared with the 1980 share of 62 per cent)
3. to encourage the use of renewable energy sources (in 1980 these accounted for 0.5 per cent of inputs to electricity generation)
4. to reduce to 0.7 or less the Community's energy coefficient or output elasticity of energy consumption, i.e. the ratio of the rate of growth of energy consumption to the rate of growth of real gross domestic product (GDP)
5. to pursue an energy pricing policy aimed at achieving Community energy saving objectives.

Table 4.1 EEC policy objectives on energy demand 1975 for 1985 (million tonnes of oil equivalent)

	(1) *1973 forecast for 1985*	*(2)* *1975 objective for 1985 assuming 50% import dependence*	*(3)* *1983 actual turnout*
indigenous production	640	800	516
net imports	1160	675	378
total	1800	1475	894

Sources: Columns (1) and (2), Brondel and Morton (1977).
Column (3), Eurostat.

Table 4.2 EEC policy objectives on energy supply 1975 for 1985 (million tonnes of oil equivalent)

	(1) *objective for 1985*	*(2)* *1983 actual*
coal		
production	180	143
imports	40	40
lignite and peat		
production	30	31
natural gas		
production	175–225	120
imports	95–115	48
nuclear electricity		
capacity	160–200 GW	52 GW
oil		
production	180	133
imports	540	289

Sources: Column (1), Official Journal C/153 of 9 July 1975.
Column (2), Eurostat.

The last two objectives need a little explanation, but it is worth noting first that the first two were almost fulfilled by 1983 as conservation and demand restrictions continued in the aftermath of the price rises of 1979–80: by 1983 oil dependence was down to 48 per cent and coal and nuclear energy already accounted for 74 per cent of electricity generation.

Clearly as the objectives came to be framed more realistically, in terms of actual performance, they came to be achieved more easily.

Objective 4 is rather curious because the *long-run* output elasticity of energy consumption is, in the eyes of most economists, likely to be determined by a large number of complex behavioural variables, rather than policy decisions, while the short-run elasticity is an extremely volatile and unreliable statistic.

Ray (1982) has calculated for western Europe as a whole that the short-run output elasticity of energy consumption varied from year to year from -4.4 (in 1980) to +1.80 in 1963, while its long-run value averaged over 1975–80 was already at 0.55. It is difficult therefore to see any meaningful sense in setting a value for this highly variable statistic as a policy objective. A much less volatile measure is the energy ratio or energy intensity of real GDP which measures the total energy input used in the economy each year per unit of real GDP produced. This statistic is considered again below.

Objective 5 is however of considerable interest because it emphasizes the Community's preoccupation with what has come to be called realistic energy pricing or *pricing for the rational use of energy* (RUE).

The discussion below devotes some detailed attention to this as well as to the Commission's reports on progress to date with the 1980 for 1990 objectives, but first there is one more set of EEC policy objectives to consider, those published in 1985 for 1995.

Once again the sensible approach of comparison with past performance is adopted and the Commission suggested the specific sectoral objectives:

1. a reduction of 25 per cent in the overall *energy intensity* of real GDP compared with its 1985 value
2. to maintain oil imports at less than one-third of energy consumption (the 1983 level was exactly one-third)
3. to maintain or increase the market share of natural gas
4. to maintain or increase the market share for coal with continued restructuring of the Community's coal industries
5. to ensure that oil and gas input to electricity generation was below

10 per cent by 1995 while that of nuclear power was about 40 per cent (1983 shares: 33 and 15 per cent, respectively)
6. to raise by a factor of three the use of new and renewable energies by the end of the century.

In terms of objective 1, these latest guidelines make use of the more stable and meaningful energy ratios to be described below, while the other objectives reflect the sort of conservation and oil substitution that would be expected from continuing the development of pricing for the rational use of energy.

Over the period in which these evolving sets of objectives have been rolled over, others have also emerged both in the general discussion of the International Energy Agency (IEA) and the economic summits of the major 7 OECD economies. The IEA's objectives have usually avoided the sort of specific targets set out by the EEC but have generally been in tune with the emphasis on oil substitution and rational pricing. Those of the summits (in particular Tokyo in 1979 and Venice in 1980) have tended to focus on short-term specifics such as holding oil imports at the 1978 level – something which was effortlessly achieved by market responses anyway, although the US originally wanted to count UK North Sea oil production as an import to the EEC!

With this review of objectives completed, it is necessary to consider two aspects in detail in the remainder of this chapter.

First, it is essential to have a clear idea of how conservation is to be measured and how it arises through market forces. For this an overall description of the factors determining energy consumption is needed along with some evidence about their measured importance.

Secondly, it is necessary to review what the Commission means by pricing for rational use of energy and how it relates to other possible objectives of pricing policy in the regulated fuel supply industries.

In both contexts, it is useful to draw on the considerable work done by the Commission itself on monitoring member states' performance in meeting the objectives, particularly those of 1980 for 1990, and in applying the guidelines for pricing policy.

Factors determining energy consumption in the EEC

In studying the determinants of energy consumption, economists generally isolate a select number of categories of behavioural influence. It is important, incidentally, to emphasize the distinction between the typical approach of economic science in analysing and forecasting demand, compared with the approaches adopted by, for example, engineers, political scientists, or mathematicians. The distinguishing feature of the economist's approach is that it is grounded in a *theory of how people (whether consumers or producers) behave* when trying to maximize their real well being or minimize their real costs subject to the scarcity of resources. There are numerous mechanical demand relationships used in forecasting which do not imply or contain specific theories of human behaviour; as such they have no economic content.

The simplest economic model assumes that producers and consumers in aggregate seek to minimize the total costs of achieving a target level of real national income by using the services of inputs such as capital, labour, energy and raw materials. From this starting point, the proximate determinants of the short-run demand for energy that have appeared to have statistical significance in the majority of market demand studies might fall into four categories.

The first of these is real national income or output itself representing the target standard of living or level of production whose costs of achievement are being minimized. During the cheap oil era, when the other determinants of energy demand appear to have been relatively stable in historical terms, it was customary to argue that only real national output per head affected energy consumption per head, and in extreme cases public utilities and governments even argued that national income growth would always cause an exactly equivalent growth in energy consumption. This was equivalent to assuming an energy coefficient, or output elasticity of energy consumption, equal to unity, and both the evidence cited in Ray (1982) and the adoption of the fourth of the 1980 for 1990 EEC policy objectives illustrate the limitation of this assumption.

As energy prices became much more volatile in the 1970s, attention was focused on the *energy intensity of real GDP,* i.e. the amount of energy in heat or volume terms used to produce one dollar's worth of real GDP. This ratio itself may be taken as one useful way of monitoring energy conservation.

Clearly there are many different ways in which this can be analysed, but table 4.3 sets out one widely used framework that shows the four statistically significant factors found in most studies: real income (GDP) growth, relative energy price changes, technological and demographic factors, and previous habits and levels of energy consumption.

Associated with these factors are some critical parameters, i.e. well-

Table 4.3 An energy demand framework

annual change in *energy intensity* is the sum of:

(1) annual real GDP growth	×	short-run output elasticity - 1
	plus	
(2) annual real energy price rise	×	short-run price elasticity
	plus	
(3) previous year's growth in energy consumption	×	proportion of past unadjusted disequilibrium after 1 year
	plus	

(4) annual rate of population or household increase less annual rate of technological progress in real GDP

Empirical estimates of critical parameters:*

Study	*Short-run output elasticity*	*Short-run price elasticity*	*Proportion of unadjusted disequilibrium*
IEA (1982)†	0.90	-0.16	0.60
IEA (1982)‡	0.86	-0.12	0.82
Nordhaus (1977)	1.15	-0.30	0.57
EURECA (1979)§	0.66	-0.45	0.56

* Long-run elasticities are calculated by dividing the corresponding short-run elasticity by 1 - the proportion of unadjusted disequilibrium.

† Industry sector.

‡ Residential/commercial sector.

§ The Commission's energy consumption model (EURECA) described in Dramais and Thys-Clement (1979).

established statistical estimates that measure the relative contribution of the different determinants. Attention has concentrated in particular on the *output elasticity of energy consumption,* the *price elasticity of energy consumption,*[2] and the speed with which energy users adjust their past usage to their new planned or equilibrium requirements. This last may be measured as the proportion of previous disequilibrium that remains unadjusted one year later, and by dividing one minus this latter number into measured short-term elasticities, economists can calculate the corresponding long-run elasticities. From the estimates shown in table 4.3, it is apparent for example that the IEA study implies long-run price elasticities of -0.4 and -0.67 for the industrial and residential sectors, respectively.

In measuring the effect of energy prices, it is important to deflate the index by the prices of competing and complementary commodities: labour, raw materials and capital services in the industrial sector, and other consumer goods in the residential sector.

Besides real income growth and price effects, energy consumption will be affected by population increase and household formation, and in the EEC these may have held consumption up at times of recession and high fuel prices. Offsetting this is the overall rate of technological progress in producing real GDP – technology that is energy saving as well as being more generally efficient. In turn these factors are overlaid by the tendency of consumers and producers to respond only slowly to economic stimuli so that they remain partly locked into historic energy consumption levels.

Prior to 1973, relative fuel prices were very stable, and the primary determinant of energy consumption was income growth, with a short-run elasticity of between 0.8 and 0.9. Ray (1982) estimates the annual real GDP growth of western Europe over the 1963–73 period as 4.5 per cent, and with other factors stable, this would indicate that the measured energy intensity of real GDP should have been falling at roughly 0.5 per cent each year in Europe, and this coincides almost exactly with the measured fall in energy intensity calculated by Ray for this period.

This process of slowly declining energy intensity might have continued unabated after 1973 had not fuel prices and other factors become much more volatile. These other factors can then be expected to operate in two distinct ways.

On the one hand aggregate energy intensity will improve (i.e. fall) if every sector of the economy becomes less energy intensive, with strong conservation efforts being pursued by industrial and residential consumers. This is the *intensity* effect of higher prices and technological progress.

On the other hand, higher energy prices may lead to a contraction in the demand for the output of the economy's heavily energy intensive industries (such as shipbuilding, heavy engineering or manufacturing). There was already, of course, a long-term secular trend away from manufacturing and towards the less energy intensive service industries.

This *structure* effect, by bringing about a change in the composition of real GDP away from energy intensive users to less heavy energy users can also lower the overall energy intensity of the economy.

In summary, overall energy intensity may change in one of three ways:

(a) simply due to real growth combined with an output elasticity below unity
(b) due to *intensity* effects as fuel prices rise and encourage conservation and energy saving technology
(c) due to *structure* effects as these same variables encourage a switch away from energy intensive industries.

The delays in these adjustments need to be emphasized however; table 4.3's estimates indicate that a year after an energy price rise or spurt in growth, much less than half of the planned adjustment in consumption has taken place. The IEA residential sector estimate for example suggests that it could take up to twelve years for 90 per cent of the adjustment to a rise in energy prices to be completed.

In the light of this analysis, it is useful to examine exactly how the EEC fared in its conservation effort over the post-oil shock period 1973–83. Table 4.4 sets out eight different indicators of conservation, by showing the numerical value of an index for 1983 assuming the respective 1973 value for each of them was set at 100. It is apparent that the degree of conservation varies widely according to whichever measure of conservation is adopted. Nevertheless, it is arguable that this variety of measured response is explicable in terms of the demand framework just discussed. The precise measures of each indicator are given in the notes to the table and each indicator is based on unweighted totals for the countries in EUR10. Real GDP is aggregated at the Commission's own estimates of purchasing power standards rather than exchange rates.

The most notable performance was the drop in the oil intensity of real GDP reflecting the force of the oil price shocks on both final consumers of oil products and intermediate users such as the electricity supply industries in the Community. The prominence in all of the policy objectives given to switching away from oil and towards other energy inputs has clearly shown significant results, and since the community's oil suppliers are more heavily concentrated in the private sector there have been no conflicting policy objectives to the full signalling of higher oil prices to stimulate conservation. Two trends

Table 4.4 Indicators of conservation performance in the EEC, 1973–83. 1983 values of indicator for EUR10 as an index number based on 1973 = 100

1. oil intensity of real GDP*	61
2. energy import dependence†	65
3. oil import dependence‡	73
4. oil consumption	74
5. oil dependence§	77
6. energy intensity of real GDP$	80
7. final energy consumption	92
8. primary energy consumption	95

Source: Author's calculations using Eurostat data.

* Oil consumption as a percentage of GDP at constant prices and purchasing power standards.

† Share of net imports in primary energy consumption.

‡ Share of oil imports in primary energy consumption.

§ Share of oil in primary energy consumption.

$ Primary energy consumption as a percentage of GDP at constant prices and purchasing power standards.

are conflated here: a general tendency to lower energy intensity together with a switch away from oil within the total of energy consumption.

The next two indicators – energy and oil import dependence – reflect both this element of demand conservation and a corresponding supply side effect. Just as higher oil prices curtailed demand, they also rendered commercial many of the newly discovered oil fields in the deeper waters of the North Sea. The level of oil imports (which in turn accounted for virtually all energy imports in 1973) was contracted more rapidly than oil consumption (see indicators 4 and 5) as intra-Community trade in crude oil expanded through the late 1970s.

It is indicators 6, 7 and 8 which, if any, reflect energy conservation on a broad scale, as opposed to demand and supply interaction in the European oil markets. Figure 4.1 supplements the energy intensity indicator by picturing its yearly performance from 1972 onwards, and showing alongside an index of the real price of energy to final users in the major 7 OECD economies.

As the earlier discussion suggested, an improvement in energy intensity over the period would have been expected anyway on historical trends, even if real energy prices had remained constant

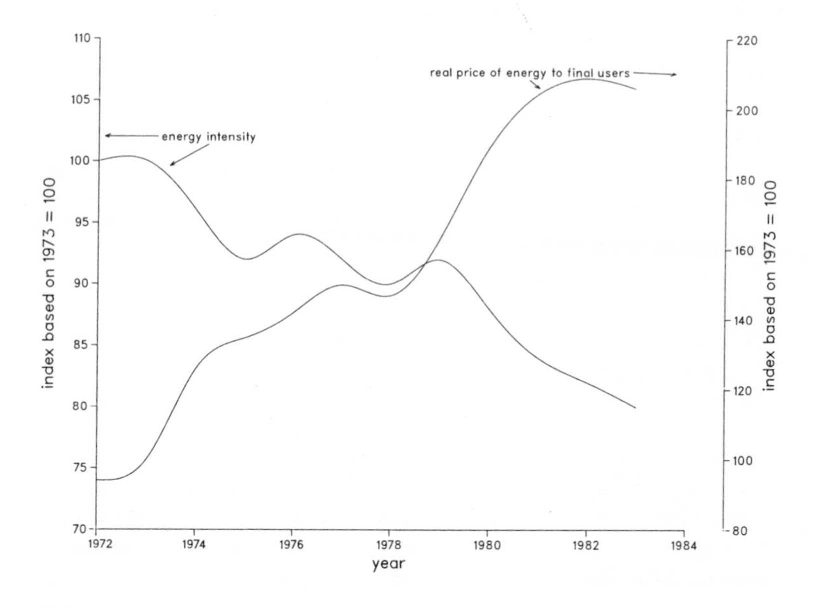

Figure 4.1 Energy intensity and real final energy prices

Source: IEA

– simply due to the effect of real income growth combined with an output elasticity of less than unity, not to say other factors such as technological progress. Nevertheless the rate of improvement over 1973–83, at about 2.2 per cent per year, is almost five times as rapid as that observed for western Europe as a whole in the preceding decade.

It is possible to detect a significant price response at work here, since the improvement in energy intensity was not smooth but had two pronounced downward shifts in 1973–5 and 1979–83, the second apparently being both more permanent and larger. Of the total fall in energy intensity of 20 per cent over the decade, three-fifths occurred after 1979. Both of these large falls followed an oil shock, and the second followed the 1979 oil shock whose price signals were not muted by deliberate macroeconomic policies to inflate the Community economy as a whole, as had occurred in 1973. A detailed study (EC Commission 1983b) of the period has suggested that this improvement very largely arose through the intensity rather than the structure effect described earlier.

It is this improvement which is put forward as a continuing aim in the first of the EEC's 1985 for 1995 energy policy guidelines, i.e. to further reduce energy intensity by 25 per cent. This requires a yearly improvement of nearly 3 per cent, so that it is in excess of anything experienced so far. This suggests that the Community's policy makers must be adopting rather particular expectations about energy demand responses over the decade following 1985:

- higher price elasticities of energy, perhaps as consumers become more conscious of energy saving
- improvements in technological progress in energy usage that significantly outweigh population and household increases as well as the slowness of consumers to adjust
- rises in real energy prices on a scale at least equal to if not greater than that experienced between 1973 and 1983.

The last two indicators which refer to the levels of energy consumption emphasize just how difficult and dilatory this conservation may be. Consumption *levels* have shown the least improvement of all indicators reflecting both inelastic price responses in the short run and the importance of previous high energy usage in locking consumers into slow adjustment to new economic factors. These last two indicators may in fact be the most accurate even if most simple measures of conservation success. Nevertheless, the fact remains that they do indicate some degree of price response which policy makers and commentators had, prior to 1973, discounted as mere economic theorizing.

It is necessary therefore to consider both how member states have implemented EEC policy objectives in general, and the role played by energy prices.

Policies of member states

In 1984, the Commission completed a substantial review of member states' energy policies in terms of the 1980 for 1990 guidelines.[3] It noted the improvement in overall energy intensity (and the associated improvement in the energy intensity of final users' consumption) and suggested that for the period after 1979 this reflected genuine improvements in sectoral energy intensities rather than changes in real GDP structure or overall economic activity. At the Community level, the 1980 for 1990 guidelines would clearly be met, but at a lower level of energy demand than had initially been forecast before the second oil shock had been evaluated.

The amount of conservation effort varied between member states with energy intensity falling most notably in Belgium, Denmark, France and Luxembourg. Italy and the UK showed the least improvement, and, along with Ireland, these countries had both the highest primary and the highest final energy intensities of real GDP in each sector of the economy.

In this context, the Commission noted the wide divergence in the decision making and control frameworks available to member states in their respective energy sectors.

Traditionally the most *dirigiste* is France where there always has been a long tradition of state intervention in energy supply. Apart from the nationalized suppliers – Electricité de France, Gaz de France and Charbonnages de France – there are many state agencies involved in purchasing and distributing resources, overseeing the nuclear fuel cycle, manufacturing generating equipment and so on.

Similarly, in the UK, the government has major powers of intervention through the nationalized fuel supply industries: the National Coal Board, the Central Electricity Generating Board, the British Gas Corporation, and so on. This enormous potential for state direction has been, however, curtailed in several ways. For many years, the UK government adopted the *Morrisonian* principle of arm's length relationships, so that the government limited itself to setting a pricing policy and overseeing the financial limits on the industries' borrowing from the private capital markets through official debt sales. The effect was that despite what the Commission saw as an enviable framework of control, intervention was qualified by a general philosophy of using only pricing policy as an instrument, in contrast to the much more extensive direct regulation used in France.

After 1979, and possibly associated with the macroeconomic response to the second oil shock, two conflicting trends emerged in UK direction of fuel supply. On the one hand, the Conservative government explicitly rejected the arm's length principle and allowed the possibility of much greater direct intervention in nationalized industry investment policy especially in the energy sector. On the other hand, it readily embraced the idea of 'privatization ' of publicly owned fuel supply utilities. Privatization in the UK context is defined as selling at least 50 per cent of the shares in a state-owned corporation to private shareholders. It should not be confused either with the idea of 'deregulation' as it applied in the context of regulated US utilities, or with the idea of increased competition from new entrants to the industry. Although UK legislation envisaged the possible evolution of privately owned secondary fuel producers paying to use the national electricity and gas distribution networks, the privatization policy was assumed to be compatible with the existence, for example, of a single, privately owned (though mildly regulated) monopolistic supplier of gas.

In Denmark, another organizational model – cooperative production of fuel supplies at municipal and regional level – could be found. The private, regionally dispersed organization of fuel supplies has been most notably characteristic of West Germany, with private fuel suppliers at arm's length from government. The oil products market was particularly free in this sense in the 1970s. Most energy market analysis emanates from privately owned fuel utilities, though there is some oversight of, for example, domestic electricity tariffs by the *Bundeskartellamt* (Cartel Office). Most local municipalities also are major shareholders in the private fuel supply utilities, and in the electricity supply industry there are several hundred different undertakings covering generation, distribution and so on. Gas, however, is different, since the main source of supply is through imports and Ruhrgas accounts for substantially more than half of these contracts.

Partly because of this difference in control frameworks, the Community's guidelines for energy policy have been implemented in different ways in different member states.

In Belgium the notable success seems to have been the reduced dependence on oil (see table 4.5) and an extensive commitment to making nuclear power the largest fuel input to electricity generation.

The Belgian government has given some financial support to energy saving schemes, but has largely regulated energy prices to cover historic costs. The chief supply problems remaining are the excessively expensive use of coal and natural gas. Belgian coal remains the most costly in the Community, and the government has retreated from its wish to stabilize output at the level of the early 1980s. In the case of natural gas, the contracted supplies seemed to be in excess of demand in the early 1980s and the surplus was being burnt in power stations. One problem with these contracts, signed at a time when energy demand projections were higher, is that their price was related to oil prices – and in the case of Algerian supplies, to crude oil prices. Since then lower cost gas contracts have become readily available throughout the Community.

Table 4.5 Oil dependence in the EEC EUR10

	Share of oil in 1983 in primary energy consumption (%)	*Change since 1973 in percentage oil dependence*
Belgium	47	-15
Denmark	65	-24
Germany	44	-12
Greece	69	-12
France	50	-12
Ireland	52	-27
Italy	67	-12
Luxembourg	35	-2
Netherlands	46	-10
UK	38	-12

Source: Eurostat

Denmark has the second highest switch away from oil dependence over the 1973–83 period, and this has largely been due to replacement by coal imports with Australia and eastern Europe as the main steam coal sources. The policy context in which this switch has occurred has featured high marginal cost based energy prices with substantial fuel taxes and the organization of district heating using excess heat from power stations. The Commission felt that both energy planning and pricing based on the rational use of energy had been very successful. For the future, supplies of offshore oil and particularly natural gas are likely to expand as Denmark remains politically divided on the use of nuclear power. Overall there has been extensive use of mandatory conservation standards, government financing of energy savings, and demand restricting pricing policy.

In Germany, the evolution of energy supply and demand responses has largely followed market forces, with the result that it has the lowest energy intensity amongst final users in the Community. In 1973 Germany was already less reliant on oil than most other member states, partly for historical reasons as the use of synthesized oil from coal had been developed in the autarkic German economy of the 1930s and 1940s. By the 1980s the rundown of the German hard coal industry remained as a primary social problem. There had been considerable rationalization among the coal producers, but for political reasons, there has always been considerable government intervention in mitigating the coal industry's decline. The electricity supply utilities have been persuaded to establish guaranteed coal contracts to power stations until 1995, although increased imports are permitted from 1988. In addition a specific element in the price of electricity to consumers amounting to 4.5 per cent, the *Kohlepfennig*, has been used to subsidize domestic coal prices. Since the coal contracts were largely signed at a time of higher demand forecasts, electricity prices have had to rise in order to cover unit costs based on lower usage. In addition, in 1983 the federal government introduced more stringent power station emission controls which have added a further 6 per cent to average electricity prices. Electricity costs may have been further inflated by the heavily regionalized structure of electricity supply which, the Commission argued, had prevented the development of economies of interconnection – such as characterize the supply networks in France and the UK.

The projected supply base in Germany looks to rely more heavily on nuclear power and natural gas, though both have run into political controversy. Nuclear power united many political opponents in sympathy with the Green movement and only in 1982 did supply prospects rise again when the federal government largely streamlined the procedure for evaluating new nuclear installations. Natural gas imports have been a source of political controversy because Ruhrgas has led the way in negotiating, at very favourable prices, long-term contracts with the USSR rather than the Community's North Sea producers.

France led the movement from coal to oil in electricity generation in the 1960s, and in the 1970s has led the switch from oil to nuclear-powered generation. In general, the Commission regarded French centralized energy planning as a major success and it is certainly true that France has pioneered the idea of nationalized industries overseeing all fuel supply with a largely marginal cost based pricing policy. In addition, energy saving has received financial incentives through specific taxes on motor fuel, tax rebates and loans on favourable terms.

The nuclear balance has left France with organizational problems however. The nuclear installations are the work of nationalized companies and large order programmes to develop their capacity and expertise together with falling energy demand led in the 1980s to surplus generating capacity. Electricité de France was virtually unique in using nuclear power generation for other than base load (off peak) supplies, so that not all nuclear capacity was being run at 100 per cent load factor. To maximize the use of this capacity, sales at prices based on short-run marginal cost can be used (i.e. ignoring the sunk element of past capital costs), so that increased penetration of industrial markets by electricity is achieved. However such a policy has left Electricité de France with a financial deficit to recover, and exports of electricity are one possible way out. During the 1960s there was some cross-channel electricity trade between France and the UK and this is now reviving, with new contracts to supply the UK being discussed in 1985–6.

Italy appears to have made the slowest progress in adjusting to the oil shocks. Oil dependence is still very high, and the only fuel to substitute for oil in the 1970s was natural gas imported chiefly from Algeria for which the suppliers sought price parity with crude oil, even to the extent of disrupting supplies in 1981. In general energy prices have been held down as an instrument of anti-inflation policy and economic growth has been encouraged on the basis of oil and gas imports. The newest gas contracts are with the USSR, Italy's largest supplier by 1986. Different Italian governments have planned to switch electricity production to coal and oil-fired generation and Community support has been given for improving the ability of Italian ports to handle coal shipments, but the plans remained unrealized by 1985.

The Netherlands energy economy has been and is dominated by its natural gas reserves, and the size of these has militated against the substitution of coal and the politically unpopular use of nuclear power in electricity generation. Although expansion of coal imports has been planned, an upward revision of natural gas reserves together with falling gas demand in the early 1980s encouraged the Dutch to reschedule coal burning and use natural gas in electricity generation. The Commission deplored this short-term use of a premium fuel, though it welcomed the possibility of renewed Dutch gas exports to the rest of the Community. These had been curtailed in the 1970s causing gas importers like Belgium to revert to highly priced Norwegian and Algerian contracts.

The UK has been regarded by the Commission as having uniquely favourable circumstances. In addition to the nationalized industry control framework with its commitment to long-run marginal cost pricing, the UK has a very diversified energy structure being quantitatively self-sufficient in coal and oil, a pioneer of nuclear development, and a large-scale gas producer. Nevertheless, performance has not, in the Commission's view, matched up to potential. For one thing, energy prices were largely held down as part of anti-inflation policy in the 1970s and only after 1980 were domestic prices in particular raised to reflect government estimates of long-run marginal cost.

In addition, the government and the nationalized gas industry have continued to resist the Commission's argument that North Sea gas depletion should be optimized in terms of Community objectives. In the nuclear field, the UK, beginning as the first country to generate electricity from nuclear power commercially, has lost its technological and economic leaderships. Chiefly, it would appear, through following a nationalistic commitment to gas-cooled reactors which have had enormous cost overruns at a time when US designed water-cooled reactors have dominated world markets in reactor design.

By 1983, the UK government had instituted a public inquiry on the first of an ambitious programme of pressurized water reactors partly constructed by French and German suppliers.

The UK coal industry, though relatively more economic than its Community counterparts, would not with its deep-mined, high cost, output compete with supplies on the world market. For many years it was protected by import quotas and deficit grants but removal of much of this protection became government policy after 1981. The Commission reported 'a difficult process of structural adjustment is now underway in the UK coal industry and will be a continuing focus of policy attention'; in fact a fourteen-month strike occurred in 1983–4 which was responsible for the first drop in total Community energy production for many years. The strike was defeated by running down coal stocks and importing heavy fuel oil bought on the Rotterdam spot market – a feature which probably delayed marginally the impact of the third oil shock, and opened the way to large-scale rationalization of the industry. Some estimates suggested that up to 70–80 per cent of UK coal capacity had a marginal cost in 1983 above the prevailing world price of coal.

The remaining members of the Community – Luxembourg, Ireland and Greece – have also reduced oil import dependence as an aim. Luxembourg whose total consumption is small in Community terms has had a successful energy saving programme, while Ireland has largely substituted newly found indigenous natural gas for much of its past oil imports. Greece is still, to all intents, a developing economy with a high dependence on imported oil which will not easily be reduced.

Pricing policy for rational use of energy

In analysing the conservation responses of energy users in Europe to the oil price rises, two initial considerations need to be borne in mind. On the one hand, several factors operate simultaneously to determine energy usage, and these include the level of economic activity, the relative prices of fuels, the long-term developments in both technological efficiency of fuel use and consumer preferences for fuel-using applicances, as well as such demographic influences as population and household formation. Separating out the conservation due to price shocks may therefore be problematic at least. On the other hand, oil price rises may only imperfectly be passed through to prices of other fuels, and then perhaps only after a significant delay. This was particularly noticeable in the aftermath of the first oil shock when the general inflation initially caused fuel prices to fall in relative terms.

This second consideration is examined first in an attempt to measure how relative fuel prices in general responded to the oil shocks. Partly this will reflect the extent to which fuel prices are administratively determined. In virtually all European fuel markets (even West Germany's where administered prices are *de facto* rather than *de jure*) there is considerable government direction of fuel pricing policy. This is particularly the case in the largely publicly owned gas and electricity supply industries. The nature of their regulation therefore often makes it unclear how these industries' prices to final consumers reflect the marginal costs of their fuel inputs or fuel substitutes, particularly the market prices of oil products. It is not unusual for public sector gas and electricity prices to reflect macroeconomic objectives such as anti-inflation policy, or central government financial targets, or to be used to subsidize the consumption of specific groups of consumers (e.g. household gas consumers in the UK in the 1970s) or the consumption of particular fuel sources (e.g. nuclear-generated electricity in France and coal-fired electricity generation in West Germany in the 1980s).

In this context the EC Commission has strongly advocated pricing for the *rational use of energy* (RUE), i.e. essentially relating the price of fuels to the marginal costs of their production in each time period, a doctrine originally advocated for the Central Electricity Generating Board in the UK, and for Electricité de France in the 1960s. (Indeed the principal architect of the modern economic theory of marginal cost pricing is the current head of Electricité de France, Marcel Boiteux.)

The guidelines for energy pricing are annexed to several of the statements about policy objectives, most notably in the annex titled 'Guidelines for a basic energy saving programme recommended to every member state: A: Energy pricing' (Official Journal C/149 of 18 June 1980). It reads

> Energy pricing should be based on the following principles:
> - consumer prices should reflect representative conditions on the world market, taking account of longer term trends;
> - one of the factors determining consumer prices should be the cost of replacing and developing energy resources;
> - energy prices on the market should be characterized by the greatest possible degree of transparency.
>
> Publicity about energy prices and the cost to the consumer of energy used by appliances and installations should be as widespread as possible.

Other statements have emphasized the ideas of prices sending reliable signals to consumers about the long-run costs of developing new energy supplies in the world market. All these statements are fully consistent with EEC competition policy and the emphasis on transparency reflects the difficulties the Commission faced in its 1975 investigation of oil company behaviour during the first oil shock.

In broader terms, the objective of signalling the long-run world market price of finding additional energy reflects the idea that economists have associated with the optimal allocation of resources, i.e. setting prices at the level of long-run marginal cost of supply (LRMC). It is fair to say therefore that the Community's ideas about rational energy pricing correspond to the doctrine of *marginal cost pricing.*

To clarify the idea of rational or marginal cost pricing, consider the simple case of the use of additional coal output either to fuel more electricity generation or to contribute to additional steel input. To derive a *socially optimal* basis for their arguments, economists may adopt the value judgement that the value placed by society as a whole on an additional kWh of electricity or tonne of steel is represented by the market price of electricity or steel:

social value of one more kWh of electricity production	=	price per kWh
social value of one more tonne of steel	=	price per tonne

Suppose now that society is contemplating the reallocation of one tonne of coal from steel production to electricity generation. The gain to society is:

price of electricity × extra kWh generated per extra tonne of coal

while the loss is

price of steel × lost tonnes of steel per tonne of coal withdrawn

Table 4.6 The basis of rational energy pricing in EC Commission objectives

efficient allocation of resources requires:

price of electricity × additional kWh per additional tonne of coal
= price of steel × additional tonnes of steel per additional tonne of coal
= price of coal

so that:

price of electricity = price of coal × additional coal requirements per additional kWh generated
= marginal cost of electricity

and, price of steel = price of coal × additional coal requirements per additional tonne of steel produced
= marginal cost of steel

and one stage further back:

price of coal = cost per man-hour × additional man-hours required per additional tonne of coal mined
= marginal cost of coal

An efficient allocation of resources requires that these marginal gains and losses are just equal, for if they are not reallocation should proceed until the falling marginal productivity of additional coal in electricity and the rising marginal productivity of the coal left in steel production bring them into balance. For the last unit of electricity generation or steel production to be worthwhile, each of the above terms must at the same time just cover the price per tonne of coal used, as shown in table 4.6.

Rearranging the logic of this last proposition leads to the contention that, e.g., the price of electricity should equal the price per tonne of coal times the additional coal requirements for generation of each extra kWh of electricity. This is the marginal cost of electricity (with similar arguments in the case of steel and coal mining itself) which leads to the general proposition that efficient resource allocation requires that prices should equal marginal costs throughout the economy.

For many commodities, economists may rely on the assumption that competitive market forces will ensure that even profit maximizing firms are, over the long run, unable to charge prices in excess of marginal cost, thereby ensuring with the aid of Adam Smith's invisible hand that private greed will nevertheless establish social benefit in the form of efficient resource allocation.

In the energy sector, such competitive forces are, however, often absent. They were clearly absent in the historical bouts of cartelization that have characterized the supply of oil and oil products, and they are often deliberately suppressed by the granting of statutory monopoly powers to public utilities in the supply of gas, coal and electricity. Even privatization of such utilities may not establish the necessary competitive forces that tend toward marginal cost pricing if these industries are natural monopolies, a proposition for which there is considerable evidence.

The pursuit of rational pricing in these terms must then rely on the regulatory process in the European fuel industries. This is bedevilled with difficulties. Public administrators may not wish to pursue marginal cost pricing for perfectly valid reasons concerning the adoption of other political objectives or value judgements. The public utilities have no competitive pressures to ensure that their marginal costs accurately reflect the most cost effective way of organizing their own operations and expenses. The central government may not wish to accept the actual accounting profits (or losses) associated with marginal cost pricing rather than historic cost pricing.

Recognizing this last factor, the Commission further advocated that in all the public sector energy suppliers, prices should at least cover average costs (as well as reflecting marginal costs). In industries that exhibit decreasing returns to scale, marginal cost pricing will ensure this happens anyway. For example reflection of long-run marginal costs in natural gas supply in both the Netherlands and the UK has, in the past, led to massive accounting profits in the public gas supply industries.

Such profits reflect the fact that in this extractive industry, setting price equal to the cost of new gas from deep water fields in the UK, Dutch and Norwegian sectors of the North Sea, leaves a substantial margin of profit over the historic costs of the initial coastal fields still under production.

Nevertheless setting aside the case of the rising marginal cost of new gas, the public fuel supply industries are classic examples of increasing returns to scale industries with natural monopoly characteristics. This is particularly evident when their distribution and transmission networks are included in the analysis. In such industries, marginal cost pricing of final supply may push revenues below the level of historic costs (e.g. of amortizing the national transmission system). It is partly this factor that has caused the Commission to emphasize that historic or average cost must be the minimum level below which prices must not fall.

In particular, the pricing policy is designed to exclude cross subsidization from activities other than energy supply or from one group of consumers to another, and prevent subsidization of alternative fuels when it is felt that a security premium attaches to oil imports. The response advocated for the latter is the oil import tax discussed above.

In particular industries the Commission felt that several problems arose.[4]

In oil the basic measure of world market LRMC is the OPEC price plus c.i.f. costs of delivery to market. But this signal has been overlaid with government controls and lack of competition amongst the distributors of petroleum products. The Commission suggested that member states should attempt to ensure the same product was sold at the same price before tax throughout the Community and figures 3.5 and 3.6 suggest some success in this. Over the 1978–84 period both industrial and domestic real oil prices rose by 70–80 per cent, well in excess of other fuel price rises. In the transport sector the real rise was about 25 per cent, mainly in diesel fuel rather than motor spirit.

In coal supply, Community production costs at the margin generally exceed import prices available on the world market, and the Rome Treaty included the objective of aligning Community coal prices with world prices. By 1985 the Commission was prepared to argue for the removal of coal subsidies by member governments in order to stimulate a continued shift towards the most economic form of coal usage. Nevertheless, the dominant coal suppliers, the UK and West Germany have remained anxious to control the rundown of their domestic hard coal industries themselves. Over the 1978–84 period, the Commission found that coal prices to industry remained stable in real terms, and were falling in 1984 due to international competition.

The Community adopted in its 1980 for 1990 guidelines, specific recommendations about electricity tariffs. These are notoriously complicated and utilities frequently discriminate amongst consumers on the basis of the use to which the power is put, and the volume of annual load, so that tariffs decline with consumption. The dominant factor in determining marginal costs of electricity load is the timing of demand since output cannot be stored and capacity has to be sufficient to meet the load in the peak period. Marginal cost pricing in electricity supply therefore implies *time-of-use* pricing with separate rates for day and night, weekday and weekend, summer and winter and so on. Progress towards time-of-use pricing, which has a long tradition in the UK (in the Bulk Supply Tariff of the Central Electricity Generating Board) and in France (in the *tariff vert* of Electricité de France) has been limited in the rest of the Community. So substantial, however, have been the load management improvements in France and the UK, that there is widespread pressure on utilities in all parts of the world to take seriously the design of time-of-use tariffs. Since the cost savings from shifting load from peaks to a more even distribution are particularly noticeable when generation is by nuclear power, it is likely that the switch to nuclear and the movement towards time-based tariffs will go hand in hand. On average, industrial users faced higher real price rises than residential users over the 1978–84 period, but the real rises themselves at about 15–20 per cent were moderated by the switch from oil to coal and nuclear power.

Natural gas pricing has been a perennial problem in energy supply controversies and is developed more fully in the next chapter. At this point, it is worth noting that the long-run marginal cost is, by implication, given by the opportunity cost of substitute fuels. In certain premium uses (e.g. in ceramics and glass manufacture) this could be quite as high as the price of electricity; in general residential use the competitive fuel is heating oil, while in bulk industrial heat-raising the competitive fuel is heavy fuel oil.

In some parts of the Community gas prices are indexed to their substitutes, and the Commission estimated that over the period 1978–84, industrial gas prices rose by approximately 80 per cent in line with the rise in residual fuel oil. On the other hand, domestic prices were contained more rigorously, rising by roughly 40 per cent over the same period, only about half as much as the rise in domestic gas–oil prices.

The real price of energy to final users is the outcome of several conflicting factors. Predominant over the 1970s was the rise in crude oil prices, but as figure 4.1 shows, the measured final real energy prices shown there rose by about 100 per cent over the ten years from 1973 although real crude oil prices had risen by about 500 per cent over the same period. Among the reasons why the follow-on in final energy prices is less than the initial crude oil price rise are the facts that other inputs also contribute their share of final energy production costs,

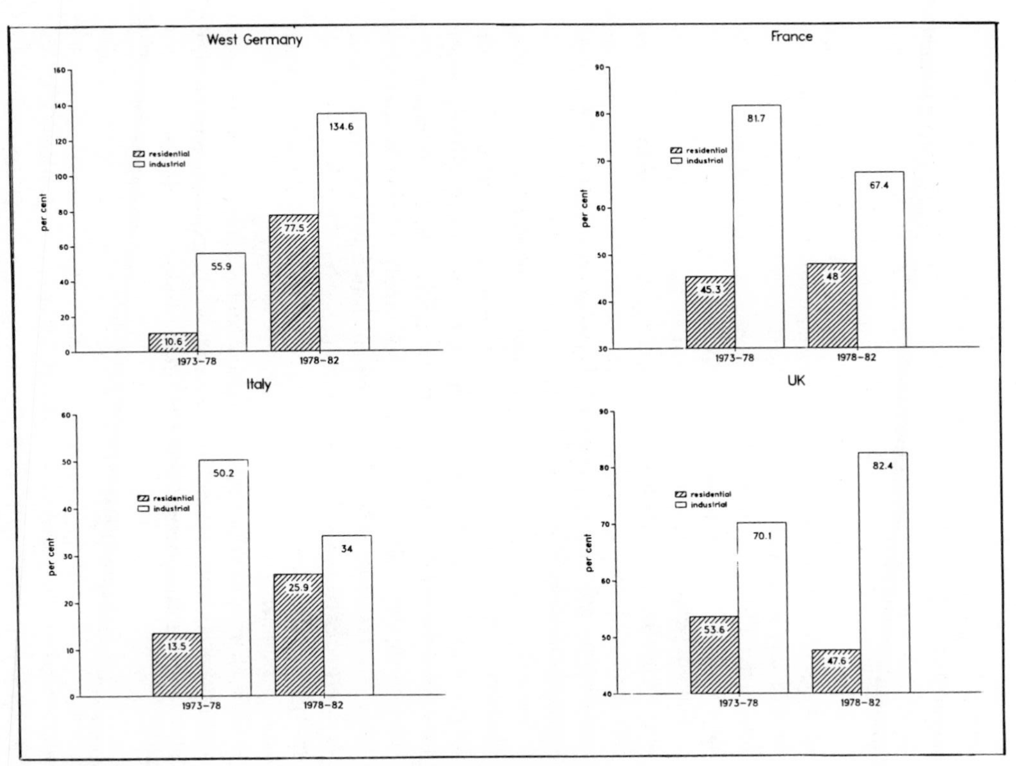

Figure 4.2 Real energy price rises for final consumers

governments may mitigate price shocks, and consumers may switch away from oil intensive production of fuels.

In terms of the aftermath of the first two oil shocks, the rises in final energy prices were about the same (47 per cent from 1973–78, and 42 per cent from 1978–82), although the first oil shock was proportionately greater. As the discussion of macroeconomic responses pointed out, the second oil shock was characterized by a containment of overall price inflation, allowing the message of higher real energy prices to be less muted in the 1978–82 period.

Within the overall rise, there were considerable national and sectoral changes. Figure 4.2 shows the percentage rises in real energy prices broken down by residential and industrial sector consumption for the four major EEC economies: West Germany, France, Italy and the UK. At this more disaggregated level, two factors stand out.

First, all four countries were united in favouring residential consumers relatively to industrial users in the aftermath of both oil shocks. To some extent therefore, objectives of equity, protection of the poorer consumers or simply electoral risk avoidance overlaid the commitment to rational energy pricing.

Secondly, there were clear national differences with West Germany's real price rises far exceeding those of Italy, especially in the second period when the commitment to rational energy pricing might have been thought to be greater. It is possible that the substantial differences in the degree to which energy supply is publicly owned in these two countries permitted the much clearer West German signals of energy price rises. The commitment in Italy to containing real energy price increases led to considerable deficits in the state-owned energy suppliers.

This chapter has focused on demand and pricing aspects of Community energy policies, and it is necessary to complete the picture with an examination of supply decisions in the following chapter.

5 Investment in European fuel supply: the policy issues

One of the predictable responses to energy price rises, and one which has characterized the 1970s and 1980s, is increased supply of indigenous fuels. In meeting the demand–supply balance, in addition to the conservation response documented in the previous chapter, five sources of additional fuel supplies can be examined: coal, oil, nuclear electricity, natural gas and 'alternatives' (i.e. solar, wind, wave, geothermal and biomass sources). Each raises important economic and political issues of its own and these are the focus of attention in this chapter.

Coal was rescued from the doldrums by the rise in oil prices but competition between increased international availability of cheap coal from the USA, Australia, South Africa and eastern Europe has still produced problems of adjustment for the expensive deep-mined coal industries of West Germany, the UK, France and Belgium. In addition, coal-burning electricity stations have increasingly been blamed for acid rain pollution in West Germany and Scandinavia.

Oil has revolutionized the economy of Norway (and helped persuade Norwegians to stay out of the EEC), and boosted consumer spending in the UK while that country's industrial output was declining. The management of oil reserves, the structural effects on the rest of the economy and the viability of investments in high-cost oil when the oil price is falling have all raised issues of real difficulty for both countries.

Following the oil price rises, those countries that had become heavily reliant on oil due to not having or switching away from coal in the 1960s saw nuclear power as a newly commercialized energy source. Both France and West Germany have attempted to embrace nuclear power on economic grounds despite considerable opposition from environmentalists. To a lesser extent this issue has also arisen elsewhere in Europe.

Natural gas has had many advocates as the cheapest and most convenient source for space and water heating in the domestic and industrial sectors. The rapid depletion of fields and penetration of fuel markets before the oil price rises necessitated serious price revisions in the 1970s and 1980s and raised questions about the availability and riskiness of future gas supplies from the USSR, Algeria and the Middle East, compounded by relationships between gas prices of regulated or state industries and world oil prices.

'Alternatives' or 'soft' energy sources are universally popular as part of national fuel strategies and a 1985 opinion poll survey for the EEC Commission's Directorate of Energy found that the Europeans sampled saw alternative energy sources as the most appropriate solution to the problems of fuel supply. Notwithstanding their popularity, they remain grossly uneconomic at current price levels for conventional fuel sources.

Nuclear energy

Though frequently presented as complex, the essential economics of evaluating nuclear power is relatively simple and follows from the ideas of pricing for rational use of energy discussed in the previous chapter. The brief summary given here appears to reflect more or less the approach used both by the Central Electricity Generating Board in the UK and by Electricité de France. Indeed, the economic evaluation technique for nuclear power is not the main issue dividing proponents and critics; rather it is the assumptions about costs and values fed in to the appraisal which distinguishes the supporters from their more well-informed critics.

The first stage in the appraisal is the calculation of a demand forecast for electricity loads for the lifetime of any contemplated nuclear investment. Since the typical 1000 MW pressurized water reactor used in the Community may take 5 or 6 years to construct, and have an economic life of 20 to 30 years after that, this forecast itself is rather problematic. It is essential however because the level of future demand will determine the marginal cost of electricity generation from existing plant over the nuclear facility's life and the latter is the essential yardstick for economic comparison. The forecast needs to take into account any provision of excess capacity required to maintain the reliability of the generating system in the face of breakdown and plant outages. In this context, it is worth noting that many European electricity systems are felt to have excessive margins of spare capacity to ensure reliability.

The economic appraisal of consumer willingness to tolerate less reliable but less costly generating systems is in its infancy, and the usual technique is to rely on engineers' rules of thumb.

With a lifetime demand forecast, it is possible to build up a forecast of what will be the marginal generating set on the system in any year of the nuclear plant's life. The marginal generating set will be whichever is on standby or about to be retired at any given time. The unit running cost of the marginal generating set then provides the marginal opportunity cost of generation. The unit running cost of the new nuclear plant (an item which itself will rise over time as the new plant ages and becomes obsolescent) can be compared with the marginal generating cost and the difference is the cost saving attributable to the nuclear plant at any point in its life. This lifetime profile of cost savings is discounted back to the present day and compared with the immediate installation cost of the new nuclear plant. The resulting difference is often known as the net effective cost (NEC) of the new plant.

It is usually the case that several investment options are available to the utility in question; e.g.

- new nuclear investment
- alternative (perhaps coalfired) investment
- retaining old capacity instead of investing

The NEC of the second option and the corresponding net avoidable cost (NAC) of the third option need to be calculated (making the appropriate economic allowance for differences in working lives of each investment option). The preferred option will be that with the lowest NEC.

If NEC is negative for any option (as the Central Electricity Generating Board, CEGB, and Electricité de France, EDF, would argue was the case in 1985 for pressurized water reactors) then it should immediately be constructed and the existing marginal generating set scrapped, because then total system costs of generation will be lower.

A complete appraisal now returns to the beginning because lower system costs and scrapping of the marginal plant should have lowered the marginal cost of electricity supply itself, which, under pricing for rational use of energy (RUE), will imply lower electricity prices and hence higher demand forecasts. The appraisal can be repeated until such marginal changes no longer occur.

The cost characteristics of nuclear power have two notable implications for the appraisal of generating systems. Of all types of thermal electricity generating equipment – nuclear, oil-fired, coal-fired, gas turbine – nuclear has the highest capacity cost of construction, but the lowest running costs when in operation. The latter fact ensures that, if built, it will be used, generally, for base load operation: i.e. run continuously throughout the year meeting basic night loads as well as peak daytime loads. (Only in a virtually all-nuclear system would some older nuclear plant be used for peak loads only. France will be approaching this point by 1990.) This would seem to imply that the cost savings from smoothing out peaks in electricity demand will be greater, the greater the concentration of nuclear power; such anyway is the argument of the Commission and many utilities. In this case a growing emphasis on time-of-use based electricity tariffs seems likely.

The second implication of the fact that nuclear power has high capacity but low running costs, compared with the alternative options, means that the appraisal is extremely sensitive to the choice of discount rate used, and this in turn may reflect the opportunity cost of borrowed capital which the utility can 'crowd out' of other, usually private sector, investment. The lower the discount rate, the more highly favoured will be investment in nuclear power.[1]

This last point, the choice of discount rate, is only one of the problems highlighted by opponents of nuclear power investment. One useful commentary (Pearce 1979) has characterized these under six headings.

1. Choice of discount rate
2. Cost of waste disposal
3. Low-level radiation emission
4. Risk of accident
5. Problems of civil liberties
6. Problems of weapons proliferation

Pearce has argued that while an economic assessment of the first four issues is feasible the last two address such deep political issues that cost–benefit analysis (cba) is rendered useless in the context of nuclear power.

Usually three objections are raised against economic cba of nuclear power:

(a) cba disregards the fact that the consequences of a nuclear programme may remain a hazard for thousands of years

(b) cba pays no explicit attention to the welfare of future generations
(c) discounting, in particular, short changes future generations by biasing choice towards projects (like nuclear power) whose clean-up costs are long delayed.

In the context of the first argument, the objective of cba is to measure all the costs and benefits of a project whenever they might occur, and if reasonable calculations of very long-lived consequences can be made then they should enter conventional cba. Nevertheless, they will have a reduced impact after being discounted. However, the point of cba is to show how, by setting aside now a part of the net present value of a programme, all of the future undiscounted costs can be met when they occur.

The second argument arises because the cba of a nuclear power programme may turn out favourable because of the weight of the net discounted benefits in the 20- or 30-year economic life of a plant. In this case the dominant factor will have been the willingness to pay for additional electricity shown by the consumption forecasts for the next twenty or thirty years. However, the consumers involved will partly determine their willingness to pay on the basis of their hopes, fears and expectations for their heirs, and the continuation through time of each generation's concern for its own children ensures that both short– and long–term projects take some account of future generations.

The third argument is wrong in assuming that only nuclear power has long delayed costs – so do many other forms of investment project. Discounting makes no assumptions about which comes first: benefit or costs, and simply reflects a view about how much extra future income or consumption is needed to compensate the current generation for sacrificing some of its own consumption possibilities in order to undertake the investment.

In this last context it is worth reflecting how the discount rate for nuclear power programmes might be chosen. If the investments were carried out by unregulated private enterprises, then they would choose as the discount rate the after tax rate of return on alternative private sector investment programmes of equivalent risk (a rate known as the social opportunity cost of capital). Because of the rates of company profits taxation in Europe and the USA, such private sector energy investments may discount the future at a real rate of anything between 10 and 20 per cent per year.

However, nuclear power programmes, especially in Europe, are usually undertaken by highly regulated or state-owned electricity and nuclear industries. In such cases, governments set out guidelines on the choice of discount rate to be used, and the Central Electricity Generating Board, Electricité de France and the other European utilities are no exception to this.

In choosing a social discount rate for public investments, economists usually argue that governments ought to take an explicit intergenerational viewpoint. Obviously the way in which this is done is a political decision, and economists' views on this have only normative (i.e. value-based) significance rather than objective or positive significance. It is true, however, that many economists argue for the adoption of a social discount rate below that used in private industry in order to avoid the distortions introduced by company taxes, private sector aversion to risk and conservative behaviour about short-term benefits and costs. In particular a substantial consensus of economists would favour choosing a social discount rate that gave equal weight to the well being of every generation into the distant future. Obviously it becomes a matter of political choice how such well being is to be imputed to generations yet unborn, but the result of such hypothetical calculations (which are unavoidable for both nuclear opponents and proponents) usually results in public sector discount rates well below private sector rates of return. This certainly seems to be true of electricity investment in Europe. Basing the social discount rate on this approach of *social time preference* rather than *social opportunity cost of capital* ensures that discounting does take future generations' welfare explicitly into account. Such a social time preference rate of discount is usually assumed to be positive (suggestions for European and US economies are around 2–3 per cent per year in real terms) because: (a) it is assumed that economic growth will lead to subsequent generations being better off *per* capita than our own generation, and (b) it is assumed that the social welfare of *extra* consumption per head falls as consumption per head rises. The first assumption is clearly a technological forecast, while the second is a value judgement which may or may not be adopted.

The second of Pearce's (1979) categories concerns the costs of waste disposal from nuclear facilities, and the essence of the objection is that even in the mid-1980s no long-term means of waste disposal had been satisfactorily developed on an economic scale. By implication therefore those conventional economic studies which were favourable to nuclear power would only incorporate the costs of temporary storage in water tanks. Since the long-term storage costs were unknown it has been argued that nuclear power ought not to be acceptable.

This objection has carried particular weight in the opposition to nuclear power in Germany, and Lucas (1985) has argued effectively that there were institutional factors in Germany that particularly favoured opposition to nuclear power on the waste disposal question. During the 1970s when there were plans for up to 50 GW of nuclear generating capacity by 1985 (the actual 1983 figure was 11 GW) it was a condition of obtaining licences to build and operate nuclear power plants that licences were also obtained for ultimate waste disposal and reprocessing. Since the potential sites were owned by the state governments through which the so-called 'citizen-lobbies' could exert pressure there was considerable room for conflict. Those in the nuclear industry tend to refer to local opposition groups disparagingly as suffering from the *nimby* syndrome: nuclear power is acceptable so long as it is *not in my backyard.* In any event local opposition groups, especially through Green Party activists, were able to use objections to waste disposal schemes to halt the utilities' switch to nuclear power – the most celebrated case being the government of Lower Saxony changing its acceptance of the Gorleben site after political opposition. By 1982 Germany had relaxed its licensing conditions sufficiently to permit separation of the operation and construction of nuclear plant from the question of ultimate waste disposal.

The third category of objection concerns the emission of low level radioactive waste from nuclear power plants and the consequent lowering of mean life expectancy in the vicinity (especially amongst nuclear plant employees) relative to the population as a whole. Two questions are at issue here. First, is the differential risk statistically significant and, though still unresolved, sufficient evidence on measured emissions had been accumulated by the mid-1980s in many European countries to make this a continuing contentious issue. Secondly, if there is a reduction in mean life expectancy, or heightened risk of some types of death, how is this to be allowed for in the cba? It is all too easy here to imagine that economics is trying to put a monetary value on human life. Such would be a futile exercise. Nevertheless it is possible to compare people's willingness to pay to avoid small extra risks, for example by comparison of house prices between areas of differential risk, or fares on transport modes of differential risk and so on. The willingness to pay to avoid very small increases in the probability of death is a well-observed statistic in road accident and other hazard surveys. The principal curiosity is the huge differences in, for example, government payments to avoid quite similar risks. Celebrated examples include decisions not to legislate for child-proof drug containers at a cost per life saved that was a tiny fraction of the cost per life saved actually incurred in reinforcing high-rise structures.[2]

The same questions arise in the fourth cba issue: major hazard and accident risks. Since the dioxin release at Seveso in Italy, most Community member states make hazard assessments legally binding on potentially dangerous installations. The bulk of opinion is that the risk of nuclear accidents is many times smaller than for other occupations or activities. John Gittus of the UK Atomic Energy Authority's Safety and Reliability Directorate makes, for example, the following comparison (Gittus 1986). While the risks of individual early death or fatal cancer for people living near a nuclear reactor that went through a degraded core accident are respectively about 2 in a billion reactor years and 1 in 10 billion reactor years, the risk of accidental death in society as a whole is 3 in 10,000 per year. More significant may be *societal risk* defined as the frequency with which an accident can occur multiplied by the total number of population health effects associated with it. For the same degraded core accident, Gittus' data suggest, for example, a societal risk of 100 or more early deaths has a frequency of between 1 in 100 million and 1 in a billion reactor years. The critical cba question is then whether the sums that could be spent on further reducing these risks could not yield greater returns in some other risk-reducing activity, e.g. wider health screening.

Attention on this aspect of nuclear power was stimulated by the accident at Three Mile Island in the USA when a widely used type of pressurized water reactor suffered a loss of coolant and vented radioactive steam.

The US President's Commission, 1979, on Three Mile Island reported that while it was the worst accident in the history of commercial nuclear power generation, it did not mean that nuclear power was inherently too dangerous as a form of power generation. The main consequences were the $1–2 billion costs of replacing the electricity generation and cleaning up the plant. The major health effect was mental stress exacerbated by delays in clarification of the nature of the accident. The physical health effects were too small to be detectable statistically. The causes of the accident were human and mechanical errors, partly reflecting poor training and preparation of control room staff, and the recommendations mainly hinged on a reorganization of the inspection authorities and training of personnel.

In April 1986 the debate on the safety aspects of nuclear power received a tragic impetus when a pressurized water reactor at Chernnobyl in the Ukraine suffered a chemical explosion and fire, releasing large amounts of radioactivity to the surrounding cities and farmland and causing a cloud of radioactivity to drift north-west over Scandinavia. The fatalities associated with this catastrophe are likely only to emerge over a number of years, but it has already been called the worst disaster in the history of nuclear power.

In the immediate aftermath of Chernobyl three issues were prominent in the public debate. In the first case, opponents of nuclear power clearly felt it gave very strong support to their case that nuclear generation of electricity is too hazardous for society to adopt. The accident itself clearly altered the frequency calculations underlying the probability of future large-scale accidents. Against this view it was argued by the nuclear industries in Europe that basic differences in reactor design, cooling systems and pressure containment vessels between the European and Soviet models limited the relevance of the Chernobyl disaster to EEC operating experience with nuclear generation of electricity. Nevertheless it is likely that valuable lessons will be available about emergency procedures used at Chernobyl and that given the paucity of accident experience in the nuclear industry, these lessons will greatly condition subsequent operating procedures. The most important consensus view that emerged in Europe was that the overriding purpose should be a co-operative effort by the world's nuclear industries to help the Soviet Union improve the safety and operating characteristics of its other nuclear plant. The EEC particularly offered help in this respect.

The second issue concerned the free exchange of information between countries about the nature of nuclear accidents. Members of the UN including the Soviet Union had agreed to the obligation to register information about such accidents immediately with the International Atomic Energy Agency, but this registering of information was one notable casualty of the Chernobyl disaster. Such information as was initially available arose only through atmospheric monitoring by countries in the west as radioactivity levels rose. Secrecy about nuclear accidents has long been characteristic of the industry – especially in the view of nuclear opponents – and had been a criticism levelled at the UK government following the 1957 fire in a reactor at Sellafield. In the aftermath of Chernobyl this Soviet reluctance to discuss or investigate the accident in public clearly did much to raise distrust of nuclear power in general.

The third issue concerned the calculation of social risk assessments to be attached to nuclear power cost benefit analysis in the EEC members' programmes. It has been noted above that such hazard assessments are carried out but do not enter as a cost of human life element in the cost benefit analysis. Following Chernobyl these procedures and the social cost penalty implicitly attached to nuclear power were likely to have a much greater impact on public reluctance to accept an economic cost benefit analysis of nuclear power. In particular the view that strictly economic assessments are unable to incorporate deep-seated social anxieties about fuel sources is likely to gain wider acceptance.

The last two categories of antinuclear argument take the issues outside the realm of economics altogether and into the arena of political philosophy. The issue of civil liberties arises because an increased emphasis on nuclear power would almost certainly require increased consciousness of security and surveillance by the state in the supply of energy. The issue of weapons proliferation arises because parts of the nuclear fuel cycle (see page 116) require technologies like uranium enrichment which render increased capacity to manufacture nuclear weapons. Conventional cba cannot meaningfully resolve such issues. In addition, there is an argument by some economists to the effect that since the focus of economics is on the narrow basis of the aggregated utility derived from personal consumption, it cannot encompass fundamental questions about the nature of society which arise in the context of major changes in technology such as nuclear power.

However it is just these issues that have been the most contentious in the opposition to more widespread use of nuclear power. The civil liberties issue stems from the fact that the raw material of the nuclear fuel cycle could be put to use in terrorist or criminal activity in order to blackmail governments through threats to local or national populations. Consequently there is no doubt that people would like to be protected from this possibility, but determining whether they would prefer this by the possibility of increased police powers in a nuclear-orientated society or by forgoing nuclear power altogether is currently one of the main issues dividing proponents and opponents of nuclear-generated electricity, since the proponents doubt whether nuclear power would in fact threaten civil liberty at all. Economists have long recognized the difficulties inherent in interpreting market responses as indicators of individual and group preferences in such areas where defence and policing issues arise. The formation of opinion and lobbying through the political process are the only means of clarifying such issues.

Table 5.1 Nuclear power in the EEC, 1983

	Belgium	*Germany*	*France*	*Italy*	*Netherlands*	*UK*	*EUR10*
installed capacity (GW)	3.5	11.1	27.2	1.3	0.5	8.4	51.9
share of electricity production (%)	45.7	17.7	48.3	3.2	5.9	17.0	22.4
share of total energy balance (%)	15.0	6.7	21.6	1.3	1.6	6.8	8.6

Source: Eurostat.

Somewhat similar problems arise under the heading of the weapons proliferation issue, but in addition the role of military use of nuclear power becomes involved. In fact both these issues along with the inadequacy of conventional cba for resolving the social impact of technological change were raised by protesters against nuclear power installations in West Germany and the UK. One serious problem is that there seems to be no adequate specialized forum for debating the issues in the context of particular nuclear installations. In 1977 there was an attempt to raise these issues in the planning inquiry into the expansion of the nuclear reprocessing facilities at Sellafield in the UK. However, the planning inspector, Mr Justice Parker, ruled that they were inadmissible, and hence their resolution has tended to go by default. The European Parliament has to some degree discussed these ideas, but its main attention has been directed towards doubts about the emissions from waste disposal facilities.

Under article 40 of the Euratom Treaty the Commission reviews nuclear power progress from time to time in the PINC (*Programme Indicatif Nucléaire pour la Communauté*) report; the second of these was published in 1972, and the third in November 1984.[3] In 1972 the nuclear industry was characterized as in its infancy while in 1984 it was regarded as fully developed. Table 5.1 gives for the six nuclear countries in EUR10 the basic data on capacity and market share in 1983, while figure 5.1 shows the rates of growth of the nuclear share in electricity generation for the EEC, Japan and the USA. In the case of the EEC, the greatest growth was in France (8 to 48 per cent) and Belgium (0.2 to 46 per cent).

The Commission noted several types of economic benefit which it

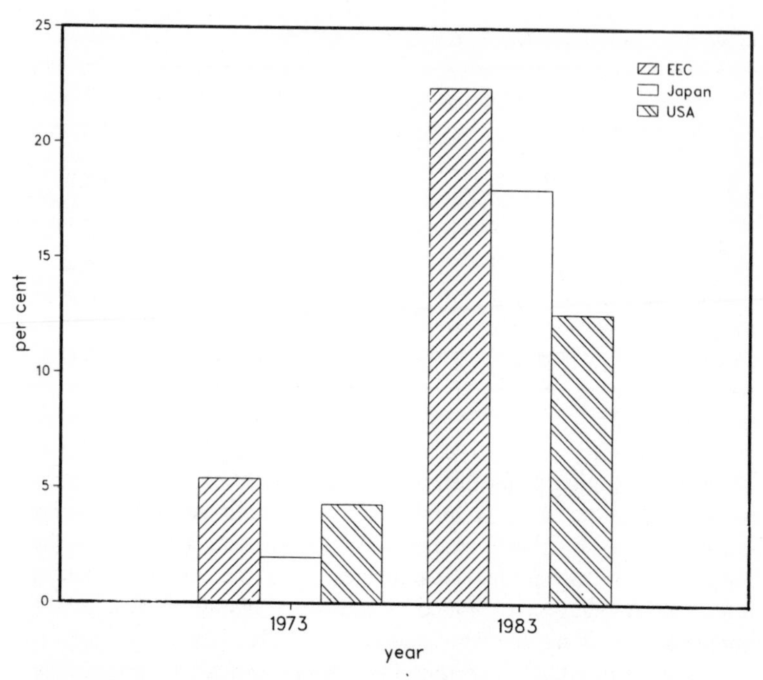

Figure 5.1 Shares of nuclear power in electricity generation, 1973–93
Source: Eurostat

associated with nuclear power. The conventional cba described above measured the primary benefits in terms of fuel cost savings, and the PINC suggests that the cost advantage of nuclear electricity measured in terms of the additional costs of generation from coal varied from 30 per cent in Italy to 88 per cent in France. Against this, it has to be said that nuclear plant construction programmes have been notorious for their cost overruns and construction delays. The UK Monopolies and Mergers Commission in 1981 even went as far as to suggest that if past experience with reactor construction was used instead of forecast cost targets the Central Electricity Generating Board's calculated nuclear cost advantage over coal would disappear.

The PINC suggests three additional, unvalued economic benefits: a reduction in the cost share of imported energy if nuclear power is expanded since although imported uranium would be used, the raw material fuel input has a much lower share of costs than is the case for coal and oil fired generation; in addition, it is argued that there would be few macroeconomic disruption costs associated with fluctuations in the world uranium price, nevertheless there is a history of cartel behaviour in the supply of uranium fuel; finally, it is suggested that there will be high technological spin offs from developing nuclear expertise. It has to be pointed out that these three arguments are traditionally part of the armoury of the nuclear industry, and they have never been successfully quantified in monetary terms.

The PINC for 1984 suggests a nuclear power objective for the Community of 40 per cent of electricity generation by 1995. This presumes annual electricity consumption growth of about 2 per cent per year with 550 TWh of the forecast 1650 TWh for EUR10 being generated in nuclear plant. Table 5.2 shows the capacity of plant scheduled to be in service by 1990, this includes plant already under construction by 1986; and to meet the proposed objective, the PINC envisages a rise in total capacity from 98 to 120 GW by 1995 in addition to a replacement of 3 GW of obsolescent nuclear plant. The distribution of the forecast capacity shows that more than half will be in France with eight new reactors, but Italy, also with eight new reactors shows the proportionately highest rate of expansion.

The PINC expects this expansion, in spite of the falling energy intensity of real GDP in the EEC, to come about through an increased electricity intensity of real GDP, already showing a 5 per cent rise over the 1973–83 period.

Table 5.2 Nuclear projections in PINC, 1984 (GW)

	Capacity in service in 1990	*Capacity in service in 1995 (after decommissioning)*	*New capacity*
Belgium	5.4	6.7	1.3
Germany	21.7	25.0	3.3
France	54.8	64.4	10.9*
Italy	3.3	10.8	8.0*
Netherlands	0.5	1.5	1.0
UK	12.5	11.6	1.1
Total	98.2	120.0	25.6

* Eight reactors in each case.
Source: PINC 1984, p. 11.

The nuclear fuel cycle begins with the supply of uranium and these projections suggest that the EEC and USA will each account for about a third of non-communist world demand by 1990. Supply is geographically concentrated and about 60 per cent of world production is under the control of five corporations, two of which: COGEMA (France) and Rio Tinto Zinc (UK) are EEC based. Following the recession of the 1970s production was cut back and there is a substantial market in already mined stocks.

The Commission expressed confidence that amongst the member states there was expertise in all aspects of the nuclear fuel cycle, often through state-owned or regulated corporations such as British Nuclear Fuels and Comhurex in France. This included operations in uranium enrichment, fuel element fabrication and nuclear fuel reprocessing. The export trade in reprocessed fuel could show considerable expansion, but the implication is that 20,000 tonnes of irradiated fuel will need to be stored in the EEC after 1990. Nuclear fuel reprocessing plants have been the focus of most protest and controversy, and their emissions record has been the most widely criticized part of the industry, especially in the UK.

The dominant reactor type used in Europe is the pressurized water reactor (PWR) initially developed under licence from US companies, but in which Framatome in France is now a leading constructor. The UK's early lead in nuclear power generation through gas-cooled reactors was dissipated by a very unsuccessful Advanced Gas-Cooled Reactor programme in the 1960s, and the currently proposed 1.1 GW expansion programme is based on the PWR partly fabricated in France.

The next stage in reactor development is, in the view of the PINC, the plutonium-using fast breeder reactor (FBR) in which the UK and the joint French–Italian–German programmes are the leading prototypes. The UK programme was heavily opposed by the Royal Commission on Environmental Pollution as entailing a high degree of uncertainty about the trend to a plutonium-based economy, so that the Superphenix of the group of three was closest to commercial size in 1985. The Commission's view is that the FBR has to be 'a fundamental component of a long-term nuclear strategy' and it has argued for strong intra-Community co-operation in financing the development of the Superphenix' successors. Such co-operation is unknown at present. By far the largest expenditure by the Community itself as a supranational body in the energy field is the research work on a thermonuclear fusion reactor, the Joint European Torus (JET). This is, in 1986, a research programme only, with no foreseeable net economic benefit likely before the next century, if ever.

Oil and gas discoveries

Among the most interesting of European energy developments has been the contribution of newly discovered oil and gas reserves to self-sufficiency and to real GDP. Three European countries have benefited particularly: the Netherlands with its huge gas discoveries of the 1950s and 1960s, the UK with first the southern basin gas and then oil discoveries in the northern basin of the North Sea, the latter coming on stream in 1975, and finally Norway with its very significant oil and gas discoveries of the 1970s.

This indigenous production of depletable energy sources has posed a variety of problems for the oil and gas exporting countries. In summary, these consist of:

(a) The rate of depletion to be adopted
(b) The method of capturing the profits for the economy as a whole through taxation
(c) The adjustment of the macroeconomy to rising oil and gas revenues.

The last issue has loomed largest for all three, and indeed has been called the problem of 'Dutch Disease'. This phrase generally is taken to represent the consequences for an economy of the supplanting of one or more leading export industries or sectors by exports of new primary products, such as oil and gas. Besides the three countries of Europe, it has afflicted Nigeria as a member of OPEC, and Australia through the latter's mineral discoveries. Sometimes the phrase is extended to encompass the type of response observed in the Netherlands – the expansion of public consumption – but this is inaccurate, since several other responses are possible, as will be seen below.

Although all three countries have explored the ideas of planned depletion policies, only the Netherlands and Norway, both of which have some tradition of central planning and consensus on policy, have actively tried to adopt depletion profiles which might not have been arrived at by market forces working alone. The Netherlands was faced with the decision in the 1960s when oil prices were at their weakest, and hence effectively put a ceiling on competitive gas prices for bulk sales into Europe's industrial markets. In such circumstances, economists often characterize the depletion decision in terms of the simple rule of thumb (the Hotelling-Solow rule) already described in the context of OPEC in chapter 3.[4]

This rule regards the stock of oil or gas reserves as an appreciating capital asset and compares the capital gains from conserving the resource with the alternative rates of return available elsewhere if the resource is depleted and the proceeds reinvested. If the real price of oil or gas is not expected to rise at a fast rate, then the annual capital gain from conserving will fall short of the alternative rate of return and so rapid depletion is warranted. In the circumstances of the late 1960s and early 1970s this rule clearly indicated the sense of rapid depletion for both Dutch and UK gas from the North Sea. As oil prices rose and the Dutch became more conservation minded in the 1970s they moved away from their old policy of low domestic gas prices below export prices, and sought to realign both prices with the export basis of the price of fuel oil.

In the UK, despite an earlier commitment by the 1974 Labour government to a planned depletion policy, the Conservative government after 1979 effectively left depletion policy to be determined by the market responses of private oil companies. Since the oil companies are used to requiring very high risk adjusted pre-tax marginal rates of return on their investments, once again rapid depletion was the outcome.

In the case of Norway, for whom the oil and gas discoveries were sufficient reason to stay out of the EEC, there was a concerted decision from the outset to plan the depletion of oil and gas in a way that would minimize the disruption and adjustments in the economy as a whole. In practice, the build up of oil production was delayed by technical difficulties and some unexpected reservoir behaviour, but even so the OECD has commented that the adjustment to rising oil revenues has still been too rapid leaving the Norwegian economy extremely vulnerable to oil price changes.

Partly, depletion, exploration and production are determined by the tax-gathering regime in operation, and the general consensus amongst informed observers seems to characterize the Norwegian regime as being harshest and the Dutch regime most lenient with the post-1983 UK system somewhere in between.[5]

The discovery of indigenous gas and oil reserves raises the problem of how governments are to capture some of the profits for the economy as a whole, but there is then a tension between maximizing the present value of the tax revenue stream and not interfering with the incentive to explore, discover and exploit the resources.

In this context, economists distinguish between taxes on the *pure profits* or *economic rent* from resource discoveries, and taxes on the *sales* or output of these discoveries.

Economic rent is a measure of the financial surplus from a project over and above all the costs of finding, developing and producing, including in the costs an allowance for the competitive rate of return to capital investment by the producer and shareholders (i.e. normal profit). If part of this economic surplus is removed there is still no alternative use of the resources including the capital tied up which could earn a higher return, and hence no disincentive to carry on supplying the oil or gas compared with a no-taxation case.

In the case of a depletable resource, the critical determinant of the depletion profile is precisely the annual percentage capital gains, i.e. the rate of growth of economic rent that appears in the Hotelling-Solow rule discussed above. Since capturing part of this growth in rent or pure profit makes no difference to supply responses, a *resource rent tax* will not cause the oil company involved to alter its market determined depletion profile. Even after handing over some of the growing rents or profits to the government there is no other depletion profile that will yield a higher present value of profit to the company and hence efficiency is maintained. Under this tax regime, since the economic rent or pure profit from a resource appreciates at the competitive rate of return on alternative investments, the annual tax proceeds per barrel of oil or therm of gas will also grow through the life of the field.

This resource rent tax can be contrasted with a sales tax in the form of a severance tax or royalty payment, by far the most common form of resource tax. Under this regime the producer (tenant) pays to the government (landlord) a fixed proportion of the selling price of a unit of output as a fee for depleting the landlord's resource. But in this case, after discounting, the present value of the royalty per barrel of oil must be lower the later the tax payment has to be made. Unlike the resource rent tax, discounting and delaying tax payments reduces their net present value. The producer therefore has an incentive to delay production start up and charge higher prices for the initial years' production compared with the no-tax or rent-tax case. In this context it may be more than coincidental that the oil field production delays in Norway were much greater than expected since Norway's tax regime has been concentrated on a royalty payment basis. Under these regimes the present value of tax payment is reduced compared with the resource rent tax.

The most substantial drawback of resource rent taxes is computation of the rents since it is (a) imperative to include in deductible costs a proper allowance for exploration expenditure, otherwise subsequent exploration effort may be discouraged, and (b) necessary to distinguish between the producer's pure economic rent and the competitive rate of return on investments necessary to keep capital resources in the sector, and which should also be a deductible cost.

In addition, it may be the case, again Norway springs to mind, that a slow start up to depletion is the objective of policy, so that in conjunction with the difficulties of rent calculation, this factor predisposes governments towards sales or royalty taxes.

Finally, in this review of possible tax regimes, it is universally the case that the actual corporate profits from all activities of the companies involved in oil field production are taxed as well. Since the design of these corporate profit taxes may incorporate elements that act like a sales or royalty tax on their oil field activities, there is already an incentive to distort the efficient depletion profile.

With these factors in mind, consider now the tax regimes in the main EEC oil- and gas-producing countries.

Norway has consistently emphasized the sales tax approach, and the major source of its oil and gas tax revenue is a progressive royalty system: i.e. the producer pays a proportion of the value of production each year, the rate rising from 8 to 16 per cent as production rises. In addition there are national and municipal taxes on company incomes and since 1975 a Special Tax on the income from oil and gas production but with additional allowances. This is generally agreed to be the harshest of the resource taxes in Europe and goes hand-in-hand with Norway's stated desire to conserve production and delay exhaustion. There are special incentives to explore in the arctic regions, but given the substantial size of the reserves and the desire to control depletion, the tax regime is not designed to accelerate production or exploration specifically.

The Netherlands tax regime for gas production has been operating since 1967 and also includes a progressive royalty structure. In addition there is a corporate profits tax and a state profit share of net profits from offshore activities.

In the UK, the initial tax regime was designed to encourage exploration and development, but following the large gains in oil prices and hence windfall profits of the 1970s, a considerable number of additional tax regulations came in, including supplementary duties, advance payments of tax, special safeguards for marginal fields and so on. The net effect of these adjustments was generally believed to have diverted resources to the Treasury at the expense of some exploration effort, and after 1983 the system was simplified. Royalties were abolished for new fields, and only two taxes remained: Petroleum Revenue Tax which seems to have partly been designed as a rent tax, and Corporation Tax on the North Sea profits of companies involved in oil and gas work, against which specific allowances to protect the development of more costly and marginal fields were possible. There is a considerable body of opinion however which argues that Petroleum Revenue Tax (a tax on profits, not revenues as its name implies) is only a very imperfect form of rent tax.

Oddly, there is an area of oil and gas production where the idea of rent tax was not only discussed but actually became the basis of policy, i.e. in UK gas production. The UK's nationalized (but privatized in 1986) British Gas Corporation was empowered to be the sole buyer of gas on the UK continental shelf, and so all North Sea gas production was settled at a price negotiated between a single, publicly owned, buyer and a small number of international oil company sellers. Exploiting its monopsony position, the British Gas Corporation set out in its price negotiations to pay no more than an amount that would allow the oil companies an 'adequate' rate of return on their North Sea investments while taking account of the growing demand for gas. In this sense, the economic rent was collected from the oil companies and presented partly to the Treasury and partly to the domestic consumer directly in the form of cheap gas.

Two difficulties with this process have been raised. On the one hand the early UK gas purchase contracts (like the early Dutch export contracts) failed to predict the rise in oil prices so that depletion was initially too rapid. The UK Treasury subsequently ordered the British Gas Corporation to pay a 'gas levy' on its resulting windfall profits. On the other hand, while many commentators are agreed that resource rent collecting was the British Gas Corporation's job, some have argued that the corporation allowed its distribution and organizational efficiency to drop so that part of the rent was dissipated by the Corporation itself.

The third and most important issue in the economics of indigenous oil and gas discoveries has been the adjustment of the rest of the economy to these developments. In all three cases the effect has been said to be very large, and not altogether welcome.

The typical European industrialized country is a heavy energy importer paying for its fuel imports by exports of, for example, manufactured products in which it has a comparative trade advantage. The discovery of oil and gas reserves is then equivalent to the gift of a new exportable commodity replacing the old leading trade sector as the source of income needed to pay for imports (which no longer need to include oil and gas). As the country switches from being an energy importer to being a leading energy exporter it loses its comparative advantage in manufactures and gains a comparative advantage in oil. There is therefore a switch of inputs, including labour, out of manufacturing and into oil. At the same time, the country's income per head has risen so that demand for all commodities including manufactures and energy will rise in proportion. This rise in demand will not in general affect world prices, but the prices of those commodities which the country produces and consumes itself without trade (construction and services, for example) will rise. This rise in indigenously determined prices relative to world prices is an appreciation of the country's real exchange rate with the rest of the world, and also represents an improvement in its terms of trade. The terms-of-trade effect in turn reinforces the improvement in *per capita* income, while the country's actual currency exchange rate will appreciate in response to the oil engendered price effects.

As the country's exchange rate appreciates, its manufacturing sector, already supplanted by energy as the source of comparative trade advantage is further eroded because foreign manufacturers become more competitive and home manufacturers less competitive. As labour and other factors are released from the contracting manufacturing sector, they can be absorbed into the expanding energy and services sectors as long as real wages adjust not only to encourage the move, but to reflect the differing labour intensity of production in the different sectors. To the extent that real wages fail to adjust (and real wage resistance is said to characterize the European economies) the level of unemployment in manufacturing will rise. Without such structural changes in the composition of GDP it is often argued that it is not possible to consume the benefits of oil and gas finds. This basic restructuring of the economy, known as 'Dutch Disease', has led to the phenomenon of deindustrialization in the UK and the Netherlands and deagriculturalization in Nigeria following the finding of oil and gas reserves.

Table 5.3 Contributions to national income from energy discoveries, 1981–3

Country	Sector share of GDP (%)	Sector share of central government receipts (%)
Norway (oil/gas)	15.0	33.0
Netherlands (gas)	7.5	14.0
UK (oil)	6.0	6.0

Source: Organization for Economic Cooperation and Development.

It is a good example of a potential improvement in a nation's well being in the sense that since income per head has risen the gainers could more than compensate the losers to welcome the investment in oil production. However, losers there will be unless there is complete mobility of all factors of production amongst all sectors of the economy, or complete price flexibility.

The precise impact of oil and gas finds has varied amongst the three countries. Table 5.3 indicates the relative importance at the beginning of the 1980s of the oil and gas sector's output contribution to national income and to general government revenues of taxes in the sector.

It is plain that both Norway and the Netherlands have preferred to tax their energy sectors more heavily in relation to the rest of the economy when compared with the UK's position. In this sense, the UK might be argued to have the most lenient tax regime for oil and gas development.

This model of the way an industrialized economy will react to oil and gas discoveries is incomplete without an analysis of the way government macroeconomic policy responds to the structural adjustments set in motion by the effect of the discoveries on exchange rates and terms of trade.

The options open to the policy makers can be seen in three categories. Since the economy is now producing more GDP from its oil and gas sector, its real income has risen relative to its absorption (spending on consumption and investment). The surplus is reflected in a healthier trade balance pushing up the exchange rate. One option is to permit an *outflow of capital* from the economy to balance the surplus of production over absorption. A second option is to permit a *rise in absorption* (spending) to use up the additional resources. The third option is to balance the increased production in energy by *reduced production of the non-energy sector* of the economy.

In practice, combinations of all three options have been at work in the three economies considered here, but in differing proportions.

The first option has been emphasized by the UK government since the build up of oil revenues after 1979. It required an economic policy that loosened exchange controls on the amount of foreign investment permitted to UK residents and companies, and a tight fiscal policy to hold down government borrowing and therefore UK interest rates relative to the higher interest rates available overseas. It has been estimated that between one-third and one-half of the North Sea revenues of the UK were matched by increased investment abroad over the period 1979–83, and this outflow has been described as immortalizing in future foreign currency earnings the finite gift of North Sea oil revenues.

It is easy to see why this was an attractive option to the Conservative government elected in the UK in 1979. The overseas investment did not add to the high inflation rate current in the UK, and the tight fiscal policy was already part of the commitment to deflate the economy and to reduce the role of the public sector – seen, on partly ideological grounds, as a drag on economic efficiency.

At the same time, the third option partly appeared as a consequence of deflating the economy. Reduced spending on the products of traditional industries, real wage resistance in those industries, and a general aim of not subsidizing the industries made uncompetitive by the advent of North Sea oil led to high rates of manufacturing unemployment. In this sense reduced non-energy production (with the factors of production consuming more leisure in the form of longer durations of unemployment) was also used to balance increased energy production.

Finally North Sea tax revenues, by obviating other tax increases that might have been necessary to finance the public sector deficit, permitted some increase in absorption in the form of consumer spending.

The Norwegian government were fully aware of the structural changes that would follow oil and gas finds, and their initial policy commitment in the 1970s was to minimize the disruptions to the rest of the economy. Partly this could be done by delayed depletion, but as the deflationary impact of recession spread from the rest of Europe, the policy option became one of raising the level of absorption. The high priority accorded to maintaining full employment persisted into the 1980s, and there was a rapid growth in consumption stimulated by subsidies and redistributive transfer payments by government. In other words, partly due to the initial good intentions to slow down structural adjustment, inflation was allowed to develop by rapidly increased government spending. This was in part a choice of what was called the *defensive* industrial policy, i.e. a deliberate decision to subsidize the traditional industries made uncompetitive by the endowment of oil and gas. There seems to be a general recognition that absorption of oil and gas incomes through public spending was too rapid, and the OECD argued that by 1982–3 the loss of competitiveness of the Norwegian economy due to subsidization of inefficiency and labour immobility had rendered it very vulnerable to the falling oil prices which were then beginning to emerge.

The use of energy discoveries to stimulate increased consumption by government transfers and subsidies has, however, particularly characterized the Netherlands, and added an extra dimension to the interpretation of the phrase Dutch Disease. The terms of trade and exchange rate effects led to the Netherlands' concentrating its export efforts on energy- and hence capital-intensive intermediate goods, which became in less and less demand as the 1970s recessions took hold. In policy terms, the gas revenues were used to increase public consumption with the OECD estimating that by 1983 government expenditure accounted for over 60 per cent of national income. A considerable amount of this went as transfer payments to households and, combined with a widespread incomes policy that reduced wage differentials through indexation, this led to what the OECD referred to as a reduced income differential between activity and non-activity. The OECD's 1983 verdict on the Netherlands was:

> Declining profitability in the non-energy sector and the over proportionate growth of the public sector (notably transfers to households) have been among the outstanding features of the last decade, and are generally seen as the root of the shrinking employment opportunities in the private economy. (OECD 1983b, p. 63)

A very simplified summary would put it like this. Norway has used its oil and gas revenues as part of a defensive industrial policy to subsidize investment in traditional industries, while the Netherlands has used its gas revenues to increase the public spending component of resource absorption. Both are felt to have rendered their economies uncompetitive and vulnerable to falling energy prices. The UK has used its oil revenues to diversify away from contracting traditional industry and to build up overseas assets and investment in new technologically biased growth industries. This has probably rendered it less vulnerable to oil price falls than might otherwise have been the case, and partly explains the UK government's refusal to support the OPEC cartel after 1982. All three countries have chosen different ways of settling what the OECD referred to as the 'delicate balance . . . between social generosity and economic efficiency' with the Netherlands and Norway favouring one side and the UK the other. The nature of the choice is, of course, a political value judgement.

European gas markets

The previous section discussed the effect of finding oil and gas on the economy as a whole, but natural gas itself has interesting economic aspects for the European economy. These include the nature of European trade in natural gas, and the dependence on large-scale cheap imports from politically strategic countries such as the USSR.[6]

It is important first, however, to underline the factors influencing the price of natural gas and the nature of demand and supply. Generally large-scale gas consumption entails the installation of capital intensive gas using equipment (storage, gas distribution systems and so on) so that buyers need the guarantee of long-term supply over the period in which these capital costs can be recovered. On the other hand, because gas is costly to transport in liquefied form, suppliers usually need to install bulk supply lines to specific market centres, and will look for the assurance of long-term contracted demands. There is therefore a concerted market pressure towards the use of long-term contracts for disposal of new gas finds. For example, many North Sea gas contracts are twenty or more years in duration. Sellers often stipulate steady supplies through the year, which buyers might then trade off at a discount on the price because their own loads vary seasonally and they have to bear the burden of storage costs, or interruptible sales contracts.

Adelman (1962) in a classic study of natural gas contracts, pointed out that this has a multiplier effect. To meet growth in annual consumption, buyers might seek further new long-term reserve contracts which makes the demand for new contracted reserves up to twenty times the annual change in consumption. In this way small consumption changes can have massive effects on the demand for new contracted reserves and equally massive effects on contract gas prices.

The delivered price of natural gas is likely to be set by the long-run marginal cost delivered to market of substitute fuels. Since natural gas is costly to transmit there may be a large distribution cost margin between the production costs of gas and, for example, of residual fuel oil. In addition, while some consumers (the non-premium market) will use gas only for large-scale heat or steam raising in which use its chief competitors are residual fuel oil or coal, others (the premium market) prefer gas for its controllability and cleanliness and may regard electricity as its closest substitute. Such uses are in particular industries like ceramics, or in household demand for cooking, space and water heating.

In looking at the European gas market, therefore, the principal factors are the growth of premium demand, the relative prices of substitute fuels, and the ability of buyers and sellers to accommodate market fluctuations in long-term contracts.

The consumption of natural gas has risen dramatically in Europe in the thirty years following the major Dutch gas discoveries. When Dutch gas was being conserved in the 1970s, the growth in consumption led major gas distributors in Germany, France and Italy to look for new contract sources in Norway, Algeria and the USSR, as the multiplier effect operated.

Figure 5.2 indicates the growth in EEC consumption and the major role of imports in supply after 1970. As table 1.1 noted, natural gas accounted for about one-fifth of EUR10 primary energy consumption by the mid-1980s. The major consuming nations are the UK, Germany, France, the Netherlands and Italy with Germany, France and Italy together accounting for more than 60 per cent of EUR10 consumption and in turn importing about 60 per cent of their consumption. Because of its high transport costs, proximity to market largely determines the choice of gas supply source, so that the major suppliers to Germany are the Netherlands and the USSR, to Italy the USSR, and for France the suppliers are Algeria, the Netherlands, the USSR and Norway. For EUR10 as a whole about one-third of gas consumption was imported in 1984, the major external source being Norway, but which is likely to be replaced by the USSR by the 1990s.

The Commission estimates that EUR10 demand in 1990 will be about 190 mtoe and one authoritative commentator, Jonathan P. Stern, has suggested that the major consumers' supply sources will be distributed as shown in Table 5.4.

There does seem to be general agreement that the overall supply–demand balance in world gas consumption does not reflect scarcity.

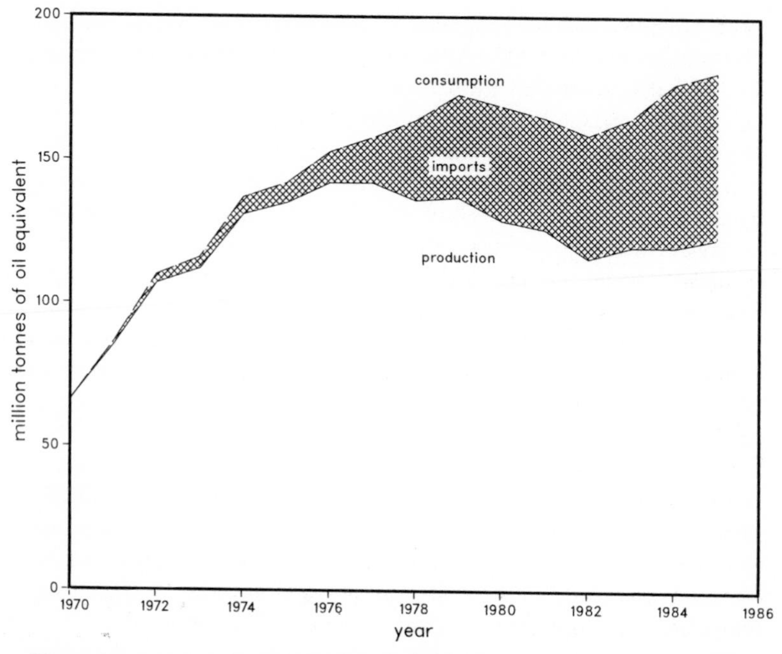

Figure 5.2 Natural gas in the EEC, 1970–85
Source: Eurostat, OECD

The International Energy Agency (IEA) (1982) study estimated that even on demand projections that were high relative to the actual experience of the early 1980s, the export potential of non-OECD suppliers would substantially exceed potential OECD gas import demand. However, more important factors are the geographical concentration of world proved reserves (the USSR has 41 per cent and OPEC 33 per cent) and non-OECD world export potential (for 1990: USSR is projected to have 37 per cent and OPEC 57 per cent).

In this context, most observers, including the EC Commission see the major gas market uncertainties as being the nature of the long-term contracts and the issue of security of supply given that the USSR will be a major source of Community imports.

The Commission points out a corollary of the long-term contract multiplier effect, that within contract flexibility and renegotiation can make for smoother responses to changing market conditions. The Economist Intelligence Unit's (1984) survey of western European

Table 5.4 Possible sources of natural gas supplies in 1990 (%)

	Consuming countries			
	Germany	*Italy*	*France*	*UK*
production and intra-EEC trade	54	34	37	77
imports from outside EUR10*	46	66	63	23
	100	100	100	100
*these are				
USSR	72	52	48	—
Norway	28	—	16	100
Algeria	—	48	36	—
	100	100	100	100

Source: Stern (1984), p. 166.

energy noted that gas contracts are particularly susceptible to disputes, and that many of the major European contracts of the early 1980s were subject to considerable renegotiation.

In this context, the least flexible supplier seems to have been Algeria which, in 1981, led a campaign to establish f.o.b. parity with crude oil prices – that is to say, importers had to pay the equivalent wellhead price of crude oil, and then bear the burden of transport costs themselves. Algeria disrupted its liquefied natural gas contracts to the USA to establish this. Eventually Distrigaz of Belgium, Gaz de France and ENI/Snam of Italy accepted this price structure and with French subsidies to Algerian supplies, long-term contracts with 'political' prices were signed, contrary to the policy stances of both the EEC and the IEA.

The USSR has proved most flexible in contracting gas supplies, and as the major supplier to Italy renegotiated a major contract at a lower price in 1984. Such renegotiation brings a spot market element into gas trade, and could lead to a general expansion in the Commission's view. In contracts with Ruhrgas and ENI/Snam, the USSR has used a price approximately indexed to the competitive delivered price of heavy fuel oil for non-premium users, and many commentators, including the Commission, believe contract flexibility and competitive pricing are likely always to weigh more heavily than supply security with large-scale purchasers. In the European context, these large-scale purchasers are effectively the major corporations in the four largest consuming countries: British Gas Corporation, Ruhrgas, Gaz de France and ENI/Snam. Nevertheless, their national governments may wish to take a different view, with a degree of political intervention in deciding long-term contracts. This possibility arises because if the USSR can be regarded primarily as a hard currency maximizer with large gas deposits to contract, it will aim for the largest market penetration, and continue to offer the best commercial terms.

This possibility in turn gave rise to a political disagreement over the potential gas supplies to western Europe by pipeline from the Urengoy natural gas fields in the USSR. In this, the USSR stayed on the sidelines, and allowed western Europe and the USA to disagree. The US view was that the European countries were becoming too dependent on USSR gas supplies, and providing a political adversary with hard currency, technology and equipment in the strategically important field of energy supply. There was US pressure on European governments not to contract for the gas and even a suggestion that Dutch gas should be sold on favourable terms to Italy to undercut the USSR source. The chief fear was that the USSR could use short-term embargo threats to enforce political demands.

In western Europe there seems generally to be less fear of contracting for USSR gas. The nature of gas transportation costs and the geographical and political obstacles to easy cartelization suggest to many that natural gas embargo fears are unfounded. It is also true that because long-term contracts are the rule, isolated embargo incidents may cause the suppliers involved to bear an irreversible risk penalty in subsequent negotiations, as has been the case with Algeria. For this reason, and also because of the large gas reserves involved, many European commentators have played down the strategic uncertainty of USSR gas contracts. In general there seems to be a tacit agreement that dependence of up to 30 per cent on USSR supplies is acceptable for any one country. This could still permit one region, e.g. Bavaria, to become much more dependent on USSR supplies; Stern quotes a possible figure of 80 per cent here. In that case, a reliable alternative transmission supply is needed for that particular region. If anything, Stern comments, the strategic dimension may allow greater probability of western Europe ceasing USSR purchases for political reasons in a world crisis rather than the converse. In this context, Kaser (1985) has emphasized that the USSR need for hard currency and steady sales in a market already well stocked with large deposits may be the principal factors at work in the 1980s and early 1990s.

A rather more serious view is taken by Maull (1981) who envisages the possibility of a double crisis of OPEC and USSR disruptions, especially if Europe's oil dependence is replaced by a subsequent OPEC gas dependency. He argues, as does Stern, that political rather than economic factors will dominate the extent of Community gas imports, and he suggests several ways in which Europe's room for political manoeuvre can be maintained in spite of USSR gas dependency. These range from cultivating a counter dependency on Community electricity exports to eastern Europe, through the build up of strategic stockpiles and alternative supply contracts, to the extension of the European gas transmission system – particularly to protect the heavily dependent parts of Germany – to include the UK, a development consistently rejected in the past.

There is considerable agreement that a further major expansion in gas consumption could be developed for the Community, and in 1984 the Commission's Director General for Energy, Christopher Audland, noted that although demand was weak in the early 1980s, in the longer term a large increase in demand was possible. To avoid over reliance on non-OECD sources, he argued that the role of the super giant Troll field off Norway would be crucial, and stated that without Troll, the Community would have to limit contracts to Algeria and the USSR (Audland 1984).

Norway and the Netherlands have used the argument that their risk-free supplies are worth a security premium (such as that discussed in chapter 3), though the Dutch did renegotiate prices downwards in the slump of the early 1980s when their reserves were revalued upwards. These lower European prices, apart from undercutting Algeria, have not been sufficient to cover the development costs of Troll, and Norway has broached the idea of EEC funding of this development to provide secure, though high-cost, long-term gas supplies.

The expanded use of gas remains a possible alternative, therefore, to the reliance on nuclear power discussed earlier. Odell (1981) indeed argued that European gas production has been more constrained by political institutions than supply availability, and that a total reliance on a much expanded indigenous production is possible and should be pursued. Without taking such an extreme view, it is possible to see that just as the relative economics of nuclear versus coal-fired generation is one possible key to European energy supply, another is the extent to which a security of supply objective is overlaid on potential gas consumption.

Coal

Western Europe has about 12–15 per cent of the world's technically and economically recoverable coal reserves (at prices prevailing at the end of the 1970s). Geological reserves are several times larger. There are four coal-producing member states in the community: Belgium, France, West Germany and the UK, but only the last two were significant by the mid-1980s. In 1983, the most recent year of data undistorted by industrial action, West German hard coal production was about 92 million tonnes and UK production about 120 million tonnes.[7]

In turn, western Europe is the world's largest coal import market; its hard coal imports in 1984 accounting for roughly 70 per cent of total interregional world coal trade of 113 million tonnes. The EEC in a normal year (1984 was not such) would account for about 57 million tonnes of this. The size of interregional world coal trade is therefore roughly equal to total UK production before 1984. In addition West Germany produces annually about 130 million tonnes of lignite which is difficult to transport and, instead of entering trade, is used to generate electricity in power stations at the mine mouth.

The reason why western Europe imports so much coal when it contains two countries with massive reserves is simply that the long-established deep-mined coal industries of West Germany and the UK cannot produce coal that is priced competitively with open-cast and strip-mined coal from the USA and Australia, and from countries where wage costs may be deliberately deflated, e.g. South Africa and eastern Europe. This fact has led to many years of painful adjustment in the Community's coal industries, with closures and redundancies at a rate only mitigated by deliberate government action.

Between 1975 and 1984 employment in the Community's coal industries fell by nearly 30 per cent, and all the industries made substantial financial losses before subsidy.

The picture had been very different in the period just after the Second World War when there was a general belief that coal would remain Europe's principal energy source for the foreseeable future. The

European Coal and Steel Community (ECSC) was the first focus of the movement towards European integration but from the outset there seems to have been a divide between the Monnet vision and the ECSC High Authority's view of its role. The emphasis became centred on the creation of a genuine common market, and the eventual alignment of European and international coal prices.[8]

As the post-war period developed and coal became increasingly ousted by oil, governments found themselves faced with the question of determining the socially acceptable rate of rundown of their coal industries. In West Germany, the 'Kohlepfennig' levy of 4.5 per cent on electricity bills was only one of a number of measures used to subsidize coal usage by electricity utilities. The utilities were persuaded to accept a long-term minimum contract take of about 40 million tonnes per year to 1995 to maintain coal's share of electricity generation. In the UK, compulsory 'coal burn' levels were imposed on the Central Electricity Generating Board, and in the early 1980s a fixed real price long-term contract for coal use was adopted. Only in 1985, after the coal strike, was a target minimum level of annual production of 120 million tonnes finally revised downwards towards the annual consumption figure of around 90 million tonnes.

Most coal trade is, like natural gas, on a long-term contract basis, but spot sales of shipments are increasing at Rotterdam and other ports. Cost comparisons are notoriously difficult given the nature of the implicit and explicit subsidies involved. Using a variety of data sources (the IEA, the UK Central Electricity Generating Board, and the Economist Intelligence Unit) it is possible to conclude that Australian steam coal delivered to the UK or Rotterdam for electricity generation had a c.i.f. price of about $48 per tonne in 1980 compared with German and UK f.o.b. prices of around $70 per tonne, while in 1983 the Australian c.i.f. price advantage over German and UK f.o.b. prices was about $50 and $20 per tonne, respectively. The Economist Intelligence Unit (1985) reports a cost advantage of imported coal over Ruhr coal of over $30 per tonne in 1985 (without specifying the import source).

Given these factors, the pertinent question in coal policy is how far should worldwide competitive forces be allowed to lead to a contraction of the European deep-mined coal industries. The question clearly has economic, political and social dimensions, coloured by the fact that in both Germany and the UK, coal has been the fulcrum of earlier economic development and is of immense historical importance. In the

UK the public debate on this issue came to a head in 1984–5 when in protest at planned pit closures the National Union of Mineworkers (NUM) went on strike. Eventually, the strike effort was called off because, although coal production dropped by over 50 per cent, electricity generation remained unaffected since power stations had built up large coal stocks, and oil-fired generation was used as a substitute. The strike led to 1984 being the first year since before the second oil shock when residual fuel oil consumption for EUR10 showed a rise.

The focus of attention in the UK and all the Community's other coal producers: France, Belgium and Germany, has been the question of what is an 'economic pit'. The economist's definition makes use of the Commission's concept of pricing for rational use of energy. For indigenous production one estimate of the long-run marginal cost (LRMC) of coal is the unit running or operating cost of the least productive pit. Arranged in ascending order of operating costs, UK and German pits in the 1980s showed a long tail of high operating cost capacity. Another measure of LRMC is the world market delivered price of coal, and an economic pit could be defined as one whose LRMC of production is below the competitive delivered world price of coal. On this criterion, considerably more than half of UK deep-mined capacity was uneconomic by 1983. A less stringent criterion was to look only at pits making a financial loss, i.e. whose average rather than marginal costs were not being recovered. This issue was clouded by the prevailing practice of banning competitive coal imports, charging domestic prices in excess of world LRMC based prices, and permitting substantial government subsidies to production costs, pension funds and welfare payments through deficit and social grants.

However, the issue of economic pits entailed several complications. In the Community's coal-producing states, the level of unemployment was sufficient to suggest that a proportion of the labour force released by closing uneconomic pits would have difficulty in finding alternative employment. This meant that the opportunity cost of mining employment (the value of output foregone in the next best alternative) might be substantially below its money wage cost, if there was no alternative to shift to. In addition it could be argued that there was a security premium that might be paid for access to indigenous coal supplies (a difficult argument to sustain in a year-long strike, however). These factors served to reduce the *shadow* costs and increase the *shadow* benefits of keeping open some of the apparently uneconomic capacity. Nevertheless, the long-run competitive market equilibrium rather clearly indicated in all four countries that to sell coal at the competitive world price, and to produce only from pits with an equivalent or lower marginal cost, entailed considerable closures of capacity and release of labour to earn elsewhere a value of output at the margin equal to its real wage costs. It was this long-run free trade position which the Community's coal suppliers had been protected from since the 1950s.

The social issues arising from pit closures are among the most contentious issues in industrial relations. As noted earlier, the location of the deep-mined coal industries in both Germany and the UK has led over the last 100 years or so to a concentration in those areas of coal-using heavy industry, and associated residential areas where the employment prospects are concentrated on the mining and coal-using industries. Permitting a rundown of coal mines therefore carries the implication of substantial disruption to working and home life and income levels of the people employed. Closing a pit can mean almost total unemployment for whole local populations; small communities may be broken up or suffer drastic income losses, leading in either case to social stress or poverty and deprivation.

Against this localized but concentrated distress has to be set the real income losses from distorting the rational pricing of energy dispersed over the economy as a whole. These can in fact be quite severe for certain groups, e.g. pensioners for whom the escalating electricity bills needed to subsidize uneconomic coal use are a heavy burden. It is tempting to ask what economics can do to resolve the dilemma, but the answer is very little. Rational use of energy merely dictates the definition of an economic pit; economics, and economists, can only stand on one side when decisions need to be taken about how an inevitable social distress is to be allocated. Economists clearly have opinions on the relative merits of distributing gains and losses but little objective analysis is possible.

It is worth noting that while economics can offer a definition of an uneconomic pit, there is nothing in economic theory to show that uneconomic pits *should* be closed. This involves trading off the hypothetical benefits to consumers and taxpayers, against the hypothetical losses to workers from redundancy and dislocation and is a value judgement, in general. However, in practice, the national governments have been offering relatively large actual cash compensation to some of the losers, permitting the possibility of turning the potential net benefits of closures into actual net benefits for the economy as a whole.

The pressure to make European coal supplies competitive on world markets, implying the closing of considerable uneconomic capacity has been very strong in the 1980s, and this was reinforced in 1985 by new Commission proposals on permitted state aids to Community coal industries.[9] Although there has been historically a consensus view on the justifiability of subsidizing the gradual rundown of European deep-mined coal capacity, in principle such subsidies contravene article 4(c) of the Treaty setting up the European Coal and Steel Community since they could distort intra-Community competition. As a result they have come under the Commission's supervision and in 1985, the Commission suggested new rules for governing such state aid to the coal industries of France, Belgium, Germany and the UK.

Following the first oil shock Community policy had been to stabilize coal production and a considerable amount of state aid was permitted. This included the deficit grants to the British National Coal Board, and the Coal Funds supported in Germany by the 4.5 per cent 'Kohlepfennig' levy on electricity bills. In 1983 state aid in the four coal-producing member states together totalled over 4 billion ECU for subsidizing current production and a further 6 billion ECU for social purposes and inherited liabilities, with Germany, France, the UK and Belgium accounting respectively for 48, 23, 19 and 10 per cent.

By 1985, the Commission felt that the expansion in world coal trade with a variety of secure supply sources meant that maintaining stable Community coal production was no longer necessary. While coal was still to be regarded as an essential substitute for imported oil, dependence on imported coal was not to be discouraged. The Commission took the view that increased competitiveness of the coal industries with less burden to the taxpayer should be the purpose. While the aids for social payments could be expected to increase as uneconomic capacity was shut down and workers retrenched or retired early, subsidies for current production should be reduced and eventually even eliminated. It sought to ensure that member states planning to subsidize the losses of indigenous coal supplies should declare an objective for their industries, and move towards increased competitiveness on world markets, qualified by consideration of social and regional problems. This was a recognition of the prevailing environment of an excess world supply of coal at marginal costs below Community prices. Even in those countries like the USA and Australia where output per man per year was nearly three times larger than in Europe, the rate of profit in coal production was only just competitive.

Apart from excess supply of competitively priced world coal, a further problem for the European coal industry was the growing concern through the 1970s and 1980s that use of coal, particularly in industrial processes and steam raising for electricity production could be the cause of the acid rain and forest die-back problems in parts of the continent. In the mid-1980s the precise nature of the problem was not entirely clear since the incidence of the pollution was slightly unpredictable, and the evidential link with coal use seemed to be circumstantial.

However, by 1984, the European Council responding to national political pressures specified authorization procedures for new industrial plant to guard against air pollution.[10] In 1983, the Commission had proposed targets to limit pollutant emission from combustion plants greater than 50 MW equivalent. It wanted member states to draw up programmes to reduce sulphur dioxide emissions by 60 per cent, and nitrogen oxide and dust emissions by 40 per cent by 1995. It reported in 1986 that little progress had been made since the differences between member states were too fundamental to achieve agreement even on the need for emission controls.

There seemed to be three broad ways of meeting the new emission standards if they could be adopted. Natural gas, a particularly clean fuel for steam raising, could be used to replace coal and lignite in power stations and oil in some industrial processes. The economic costs would reflect increased demand for an imported fuel source whose price was partly indexed to oil prices. A second possibility is the use of fuel grades with lower sulphur content, entailing in the Commission's estimate, between 4 and 8 per cent increases in electricity production costs. A final possibility is the use of 'scrubber' desulphurization units which might add about 10 per cent to the capital costs of coal-fired generating plant.

Such environmental concerns, which also became part of the 1985 energy policy objectives added more uncertainties to the projection of the Community's future energy balances.

Alternative and renewable technologies

A 1985 survey carried out for the EC Commission found the following views about alternative energy sources held by people in member states.[11]

(a) the least polluting: 1 in 2 Europeans (except UK and Ireland)

(b) the most stably priced: 1 in 5 Europeans
(c) the most appropriate solution to supply problems and energy difficulties: 1 in 2 Europeans

The IEA has classified in this group a range of sources including hydropower, solar energy, wind, wave and tidal power, biomass and geothermal energy. These are the 'renewables'. A separate category comprises the new technologies of synthesizing oil and gas from coal, enhanced oil recovery, and the use of tar sands and shale oils. This category would presumably not be regarded as 'soft' technology in the sense of the first category.

At the height of the second oil shock in 1979–80, considerable attention was focused on these new sources of energy, but the economics of their production were still uncertain. Estimates published in the press suggested that synthesizing gas from oil and coal would cost up to $55 per barrel of oil equivalent in 1980 prices while manufacture of gas or ethanol from biomass and agricultural waste might cost over $80 per barrel of oil equivalent. Solar and wind-generated electricity where these were geographically feasible appeared to have a combined generation and capital cost per kWh at least twice as high as that of conventional thermal or nuclear-generated electricity.[12]

The exception to this is hydroelectric power which has long been an important input to electricity generation in Europe. At the end of the 1970s the Community generated about 12.5 per cent of its electricity from hydroelectric schemes (and a further 0.25 per cent from geothermal energy in Italy), compared with a world share of hydropower amounting to about 21 per cent of electricity generation. France and Italy accounted for 49 and 33 per cent, respectively, of the Community's use of hydroelectricity. In general hydroelectricity shares are if anything expected to decline, and certainly in the industrialized countries of the Community the potential has already largely been exploited so that very few feasible sites remain. None of the Community's schemes is large on a world scale and only Dinorwic in Wales ranks amongst the world's largest hydroelectric facilities. (All data are from Grathwohl 1982.) Dinorwic itself is a pumping station used for peak load balancing. Nuclear-generated electricity is used at night to pump water uphill, and the rundown at peak times during the day therefore offers an immediate surge of reserve capacity at a small premium over the offpeak opportunity cost. It's capacity is about 1800 MW, roughly equivalent to a modern nuclear station.

Tidal power schemes have had some attention particularly on the western coasts of France and the UK. The first notable tidal power plant to generate electricity is La Rance at St Malo in France. However, its output of 200–300 MW is less than a fifth of the scale of modern conventional thermal plant. Despite considerable evaluation studies, tidal power schemes in the UK did not project economic rates of return even at the highest point of the oil price rises of the 1970s.

As a consequence, alternative technologies remain at the research stage with no short-term prospect of their commercial adoption, except in localized areas and in special circumstances. Nevertheless, research is continuing, and a leader in this is France where renewable energy sources are under the supervision of the Agence Française pour la Maîtrise de l'Energie (AFME). The Mitterrand plan for the energy sector proposed a renewables' output target of 10–14 mtoe and a hydropower target of 14–15 mtoe by 1990; respectively 8–12 and 12–13 per cent of total projected French primary energy production.[13]

The aim of this chapter has been to examine the uncertainties and policy issues inherent in Europe's current fuel supplies. This leads into a consideration of the likely evolution of Europe's energy balances, the topic of the next chapter.

6 The evolving energy balances of the EEC

A relatively clear picture of the conservation response to oil shocks in the EEC had emerged by the mid-1980s, and associated with this was the perception of the supply side issues discussed in the previous chapter. This foundation provided the opportunity for several exercises in analysing possible future energy market trends within Europe, and the Community in particular.

One such study is of particular interest. This is *Energy 2000: a Reference Projection and its Variants for the European Community and the World to the year 2000.*[1] This first appeared as an EC Commission Staff Paper in February 1985 and was the focus of attention in the first issue of the Energy Directorate's regular *Energy in Europe* publication.

First of all, however, it has to be emphasized that, as with many other such energy market projections, the analysis cannot be regarded as a forecast, since it cannot by definition come true. This is because the purpose of such projections is to determine whether and at what point an *ex ante* imbalance between demand and supply might occur in the energy market. *Ex post*, no such imbalance can possibly arise, because one or more of the critical variables – consumption, production, or the real price of energy which rations consumption to production – must have altered. Added to this are the exogenous shocks to the market system such as the decision by OPEC to exercise its market power in 1973–4. These are, by their nature, generally impossible to project.

Energy 2000

With a courage not daunted by the conspicuous failure of past energy forecasts by academics, governments and industrialists alike, the EC Commission drew up, between 1982 and 1985, a detailed long-run projection for EUR10 of the Community's energy consumption and supply response patterns to the end of the century. Three snapshot years were highlighted: 1990, 1995 and 2000, and assumptions were made about technological progress, the course of oil prices, economic growth and the industrial structure underlying Community gross domestic product (GDP) to cover the periods 1983–90 and 1990–2000. The reference scenario, or central assumptions, indicated that within slow overall energy consumption growth, an increase in the market share of electricity largely associated with a concentration on nuclear power would be the dominant theme. It is tempting to see this energy evolution in decades: coal for the 1950s, oil for the 1960s, conservation and gas for the 1970s and 1980s, and nuclear for the 1990s. However, the detailed projections, especially amongst different member states, are rather more complex, and the alternative scenarios also merit attention.

By the beginning of 1986 the Commission's projection, whose shortened title is simply *Energy 2000*, was going through the process of commentary by consultative committees, and had already been criticized for not emphasizing the role of coal by the European Coal and Steel Community (ECSC) Consultative Committee. The study is now considered in more detail.

Socio-economic background

The preliminary work on the projection entailed the development of forecasting models for European demand and supply. In chapter 4, the principal determinants of energy demand were discussed and in summary these can be described in the following terms.

Energy consumption, in particular energy intensity per unit of real GDP, is affected by (i) demographic change, (ii) technological progress in the economy as a whole, (iii) real economic growth, (iv) relative energy prices and (v) previous habits and levels of consumption to which consumers are partly locked in.

The Community's population, corrected for the enlargement by member states, is assumed to increase at about 0.2 per cent per year to the end of the century reflecting differing patterns of birth rate and household formation in the Community.

Offsetting the population increase will be the general technological progress evident in society at large. *Energy 2000* emphasizes that real advances in energy saving need to be embodied in new capital equipment to have any effect so that a rising level of capital investment in the Community is seen as an essential part of the process by which the technology of energy saving is disseminated. This requires that the economy of the EEC emerges from recession, and that there is real economic growth and rising capital investment. Real GDP growth of course adds to energy demand, but without real growth energy saving investment will not be stimulated, in the Commission's view. The reference projection assumes resumed growth but at a slower rate than the 1960s: 2.4 per cent per year to 1990, and 2.8 per cent per year from 1990 to 2000. Distributed over the expenditure components of real GDP, this is assumed to leave private consumption with an unchanged share, and a fall in the share of government expenditure for public consumption by about 1.5 per cent, offset by a similar rise in the share of investment expenditure. This presumes a continuation of the tendency, observed in the early 1980s, to reduce the public sector's role in national income creation.

Along with the new investments in energy saving stimulated by economic recovery, two other transmission processes are noted by which economic factors may affect energy usage. Both result from the assumptions made about real energy prices. On the one hand higher energy prices lead to reduced energy intensity in all sectors of the economy, and on the other hand, higher energy prices lead to a structural shift in the composition of real GDP away from energy intensive industries.

The reference projection assumes an initially falling US dollar price of crude oil to the Community from $30 per barrel to $27 per barrel in 1983 prices by 1990. This is due to a continuing expected weakness of demand in the world oil market. In fact by the time that *Energy 2000* was going through consultative committees in 1986, the decision by OPEC to re-establish its market share had caused the Rotterdam spot price of oil to drop below $18 per barrel, a fall of over 50 per cent in real terms. Beyond 1990, *Energy 2000* assumes rising oil prices as world oil demand recovers along with economic prosperity, until the real price in 1983 terms is $35 per barrel in the year 2000, i.e. $80 per barrel assuming annual world inflation at about 4–5 per cent. As always, the uncertainty of oil price movements can play havoc with otherwise well-designed forecasts. It is still possible that the reference projection would be borne out, but the course of oil price movements is likely to be much more volatile than *Energy 2000* presumes. Nevertheless, unexpectedly fast oil price decreases may serve to stimulate real GDP growth further so that the final projection is not wholly awry.

On the basis of the reference projection of mildly falling and then mildly rising oil prices, the shift away from energy intensive industries is assumed to continue. The GDP share of the energy intensive industries is assumed to fall from 9 per cent in 1983 to 7.7 per cent by 2000, while other industries' share falls from 25.3 per cent in 1983 to 24.9 per cent in 2000. The non-energy intensive services sector, however, is expected to raise its share of GDP from 55.5 per cent in 1983 to 58.6 per cent in 2000.

These effects depend not only on the crude oil price of course. Natural gas prices are presumed to hold up due to growth in the premium market, while coal import prices, especially from the East Coast of the USA, are expected to rise in real terms as world demand expands to take up the excess supply. Electricity prices however are assumed to fall in real terms because of the cost savings of the switch to nuclear generation.

Energy 2000 reference projection

The overall expectation is of increased energy saving per unit of GDP due both to generally higher real energy prices and to the energy saving investment stimulated by economic recovery.

With these key assumptions, the Community's forecast energy balances can be examined both in terms of gross primary energy and final consumption by end users. Figure 6.1 indicates, by comparison with the actual 1983 position, the forecasts of gross energy consumption for EUR10 for 1990 and 2000.

In the first period up to 1990 primary consumption is forecast to increase at about 1.7 per cent annually indicating a continued fall in energy intensity. The shares of natural gas and coal remain stable, but imports of both increase, in the case of gas rising by more than 60 per cent. The share of crude oil in primary consumption falls by a further 5 per cent, but the dependence on imports rises as the UK sector of the North Sea is depleted. The share of nuclear power nearly doubles from 8 to 14 per cent, and is the only relatively expanding source, apart from imports.

Beyond 1990, the effect of continued energy price rises with the application of rational use of energy (RUE) pricing and expanding investment in energy saving further reduces energy intensity, so that gross consumption is forecast only to increase at 1 per cent annually.

During this period oil again loses part of its share of consumption,

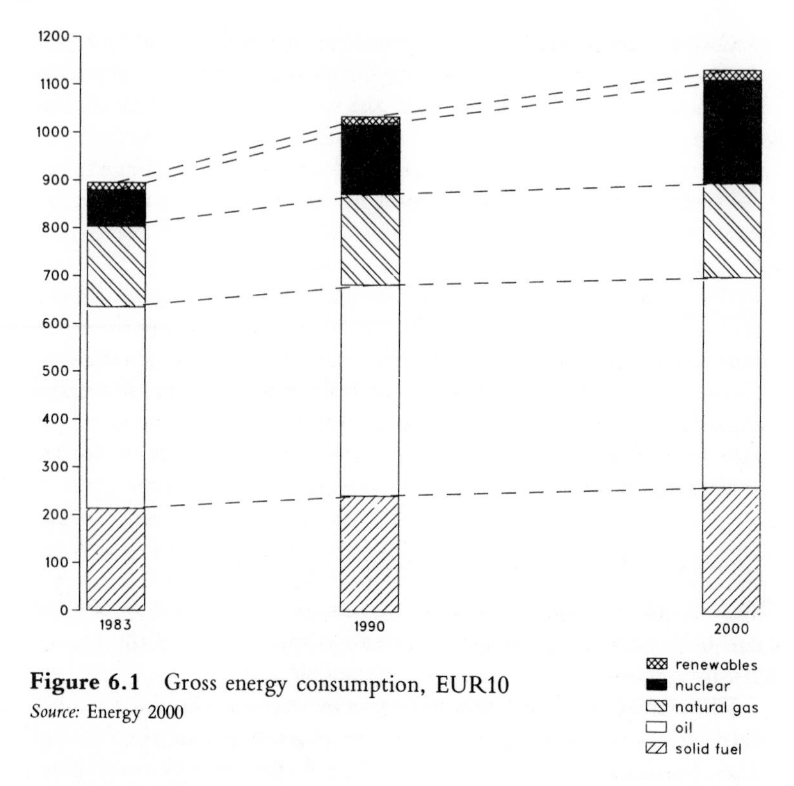

Figure 6.1 Gross energy consumption, EUR10
Source: Energy 2000

dropping to 39 per cent, but with imports staying at the same level to compensate for falling North Sea production. The market share of coal remains stable at 23 per cent but coal imports account for 35 per cent of this consumption, up from 28 per cent in 1990. With natural gas, increased levels of imports are also forecast within a slightly declining market share of consumption.

The squeezing out of oil and to a small extent natural gas is to make room for another rise in nuclear power which on this projection by 2000 accounts for nearly one-fifth of gross energy consumption (roughly equivalent to the role of nuclear power in France in 1983).

The projection therefore argues that as gross energy consumption increases with economic recovery, the supply response is through increased nuclear power and use of imports of coal, oil and natural gas. No real expansion in Community production of the fossil fuels is projected. It is worth noting that there are several commentators, e.g. Odell (1981), who would argue that EEC natural gas potential could be considerably greater.

Since nuclear power offers energy only in the form of electricity, and the reference projection assumes a changing industrial structure, *Energy 2000* must also forecast specific consumption patterns of final end users.

In terms of the broad final market sectors, very little change is expected in market shares. The share of final energy consumption by industry is expected to rise only from 29 per cent in 1983 to 31 per cent in 2000, with transport and non-energy usage maintaining absolutely level market shares. Residential and tertiary (i.e. commercial and public buildings etc.) usage is projected to lose only 2 per cent from its market share by 2000.

However, the shares of final energy consumption by secondary fuel categories are projected to change significantly, as shown by table 6.1.

In final energy consumption electricity and heat (i.e. use of district heating and combined heat and power generation) show the greatest rise and together account for over 20 per cent of final consumption. This is possible because together they are expected to gain increased shares of the industrial and residential markets: heat and electricity's share of the residential and tertiary market being projected to rise from a fifth in 1983 to one-third in 2000, with an almost similar rise in industrial usage.

This increase in the market penetration of heat and electricity is projected to come about because of the cost savings of nuclear power when compared with the reference projection's 1983 assumptions about oil, gas and coal prices. Table 6.2 indicates the projected share of nuclear power in electricity generation, and this is assumed to reach 43 per cent by 2000 (comparable to Belgium and slightly below France in 1983).

Clearly there are major uncertainties surrounding this reference projection, only one of which is the difficulty of forecasting oil prices underlined by the report's underestimate of the extent of price collapse within two years.

Alternative assumptions

One major assumption in *Energy 2000* is that economic recovery elsewhere in the world does not seriously affect the balance of supply and demand in world energy markets. Despite assumptions that the world economy might grow at 3 to 3.5 per cent a year (with particularly strong growth in the third world) the report does not envisage that world fuel supplies will be too inelastic to meet the

Table 6.1 EEC final energy consumption by fuel type for EUR 10 (%)

	1983	*1990*	*2000*
solid fuel	8	8	8
oil	54	51	47
natural gas	22	23	24
electricity and heat	16	18	21
	100	100	100

Source: Energy 2000.

Table 6.2 Fuel shares in EEC net electricity production for EUR 10 (%)

	1983	*1990*	*2000*
solid fuel	43	42	39
oil	13	5	4
natural gas	9	8	4
nuclear	22	35	43
other primary electricity	13	10	10
	100	100	100

Source: Energy 2000.

consumption growth, though clearly it does allow for some rise in world fuel prices to smooth out the rationing of demand to supply. The report's chief concern in this context is that insufficient oil and gas production capacity might be set in place in third world suppliers, in which case rising demand projections may simply lead to more substantial upward shifts in oil and gas prices.

Nevertheless, *Energy 2000* is more concerned with the importance of allowing for different assumptions in the key variables determining demand. This refers to the *economic* assumptions about GDP growth rates, and the restructuring of industry, and the *energy* assumptions about fuel prices, plant mix in electricity generation, and improvements in energy intensity.

Varying the broad economic assumptions produced little substantive differences from the reference projections. Higher GDP growth pulled up energy consumption but also stimulated the putting in place of energy saving investments, while lower economic growth reversed the directions of these offsetting trends.

One alternative scenario assumed energy intensity in each sector of GDP production improved as in the reference projection, but maintained the 1983 shares of energy intensive industries, services and so on so that no restructuring of GDP occurred as a result of slowly rising fuel prices. The share of solid fuels, which are more intensively used by heavy industry, rose, but overall energy consumption was little changed. Varying the energy assumptions had rather more effect.

Two alternative projections for oil prices are used: the low-price scenario leads to a real 1983 price of oil equal to $20 per barrel in 2000, while the high-price scenario leads to a real price of $50 per barrel in 2000.

The low oil price scenario implies solid fuels will not be competitive with oil in industrial processes, and therefore oil demand recovers towards 1973 levels with a consequent increase in imports. This scenario by implication ignores the possibility that European consumers have adjusted their subjective estimates of the probability of oil supply disruption and monopolistic price escalation since the period before the first oil shock. Even if oil prices fell in real terms to pre-1973 levels or below, it is unlikely that consumers have the same long-run expectation of oil prices that they had in the cheap oil era. Bearing in mind the 1970s' experience, oil consumers and indigenous producers are likely to have adjusted upwards their *certainty equivalent* oil price,[2] which will cause them in turn not to expand imports to pre-1973 levels nor to reduce the supply of oil substitutes to pre-1973 levels. This consideration mitigates to a certain extent the errors from projecting oil prices that are too high in the report.

The high oil price scenario returns Europe to the trends of the late 1970s with declining economic activity and energy consumption. This is the only scenario in which the Community's energy imports decline further by the year 2000. In the reference projection increased coal and gas imports make up some of the shortfall in production below rising consumption.

Energy 2000 raises one intriguing though possibly unlikely consequence of a high oil price future. Natural gas becomes a widely economic substitute for oil and given the possibility of large-scale trade expansion in natural gas, the report suggests that natural gas prices may no longer be indexed to oil prices, with the USSR leading a price fall to secure large-scale penetration of Community energy markets.

Turning to plant mix in electricity generation, there is a direct trade off between nuclear and coal uncertainties. Imposition of strict emission controls could raise costs of coal- and oil-fired generation significantly to enable nuclear power to dominate the choice of plant. Alternatively concerted antinuclear opposition such as West Germany experienced in the 1970s would open the way for coal-fired generation. The outcome may vary amongst individual member states.

The reference projection's assumption of a 20 per cent improvement in energy intensity by the year 2000 does not exhaust the technical possibilities of conservation potential especially in the domestic sector through better insulation, heat recovery, use of district heating and heat pumps and so on, but on the other hand it does assume a continuing commitment to pricing for rational use of energy (RUE). Were this to be relaxed, and energy intensity to show only a 10 per cent improvement by 2000, then the report suggests that a further 140 mtoe of primary energy would be needed, which would imply that energy imports rose once again above 50 per cent of primary consumption, and therefore approached the pre-second oil shock level of dependence.

In all scenarios, except that of the high oil price, *Energy 2000* projects a rising import dependence once again, though not in most cases anything like the pre-1973 level. To avoid this, it emphasized the dual importance of marginal cost based RUE pricing to secure a much lower energy intensity, and the emphasis in supply of nuclear power and a rising share of electricity in final consumption. The latter implies a stable role for solid fuels subject to the emission controls that are part of the Community's 1985 for 1995 energy policy objectives.[3] Natural gas is also projected to be a stable demand component but with a politically sensitive import content.

Implications for individual member states

Within the broad overall projections, differences amongst member states' relative positions did emerge.

Two member states, Denmark and the Netherlands, had set their minds against nuclear power in the 1970s with caveats appended to energy policy statements. *Energy 2000* assumed that the Netherlands does find nuclear power acceptable and that it plays a small role in primary energy consumption after 1990. Nevertheless electricity growth is forecast to be slower here than in the rest of the Community. Solid fuel use is increased in power stations but the main thrust of policy is in continuing energy savings with a repetition of the post-1973 fall in energy intensity of 20 per cent as industrial output is restructured. Denmark is assumed to continue its opposition to nuclear power, and since the scope for further energy savings is less than in the rest of the Community, energy consumption increases are covered by increased use of solid fuels and Denmark's own oil and gas reserves.

Belgium is more characteristic of the overall projection, doubling the share of nuclear power in gross energy consumption and restructuring the economy away from concentration on heavy industry. Consequently solid fuel use is projected to decline until combined heat and power schemes take off in the 1990s.

France, Germany and the UK are projected to lead the switch to nuclear power. The report refers (p. 26) to the 'breakneck expansion of nuclear capacity' in France, and forecasts for the UK that the 'marked increase in nuclear output is the most striking phenomenon ... for the rest of the century' (p. 34). All three countries are projected to achieve at least 20 per cent improvements in their energy intensity ratios by the year 2000, and the projection is 27 per cent in the case of West Germany. This is due both to a far reaching projected restructuring of the economy away from energy intensive industrial output, and substantial energy savings in all sectors. Although natural gas demand will rise in West Germany and the projection assumes continued state support for solid fuels through district heating and CHP (combined heat and power), nuclear power usage is again the main source of supply for increased primary energy consumption.

In terms of import dependence, its natural gas consumption means West Germany will continue to import substantial amounts of primary energy. However, in France dependence on natural gas and oil imports is largely offset by exports of its highly competitive electricity to Belgium, Luxembourg and through cross-channel transmission planned in 1984–5 to the UK. With North Sea oil and gas, the UK is forecast to remain energy self-sufficient for many years.

In Italy, the reduction in energy intensity at 17 per cent to 2000 is forecast to be below the weighted Community average, and a large part of its projected energy consumption increase is supplied by solid fuel and natural gas imports. Solid fuel is particularly expected to account for a massive switch from oil in electricity generation before nuclear power undergoes the rapid post-1990 expansion noted in the PINC (1984) study.[4]

Luxembourg consumption increases will continue largely to be met by fuel imports from other member states, while Ireland and Greece are projected to have the characteristics still of largely developing economies. That is to say their energy use will increase through oil and solid fuel imports.

Energy 2000: a critique

Projections like *Energy 2000*, despite being bravely constructed, have a notorious habit of being wrong, and since oil prices dropped significantly within a few months of its publication, it does not appear to be an exception.

Several more fundamental questions can be raised about the projections however, though the inability to project oil prices is an important starting point.

Any forecast of energy balances 10–15 years ahead will be sensitive to the broad underlying economic assumptions, and it is clear that the Commission has selected only a relatively narrow range of possibilities in drawing up its different projections.

In terms of economic growth the variation around the reference assumption of 2.6 per cent average to 2000 is from 1.8 per cent to 3.5 per cent (a range of 1.7 per cent). GDP growth in western Europe over the 18-year period from 1963 to 1980 in fact varied from –1.2 per cent to 6.1 per cent, a range of 7.3 per cent, though averaged over 5-year periods the range is 2.9 to 5.0 per cent (Ray 1982).

The 2000 oil price in constant 1983 dollars is allowed to vary between $20 and $50 per barrel, with the reference projection centring on $35. Over a comparable period between 1973 and 1981, the real oil price fluctuated between $5 and $35 per barrel. Although this gives the same range in absolute terms, it means that the report's assumed variation is between –33 per cent and + 66 per cent of the initial price of oil whereas the actual experience is of a variation of + 500 per cent of the initial value.

These casual observations suggest that, like many other similar studies, *Energy 2000* permits only a very narrow range of possible scenarios when compared with the range of variation observed in the recent past. It is therefore an implicit assumption of the Commission that both energy markets and national economic growth experience will be markedly less volatile over the 1985–2000 period than they were between 1970 and 1985. There is, of course, no evidence to support this implicit assumption.

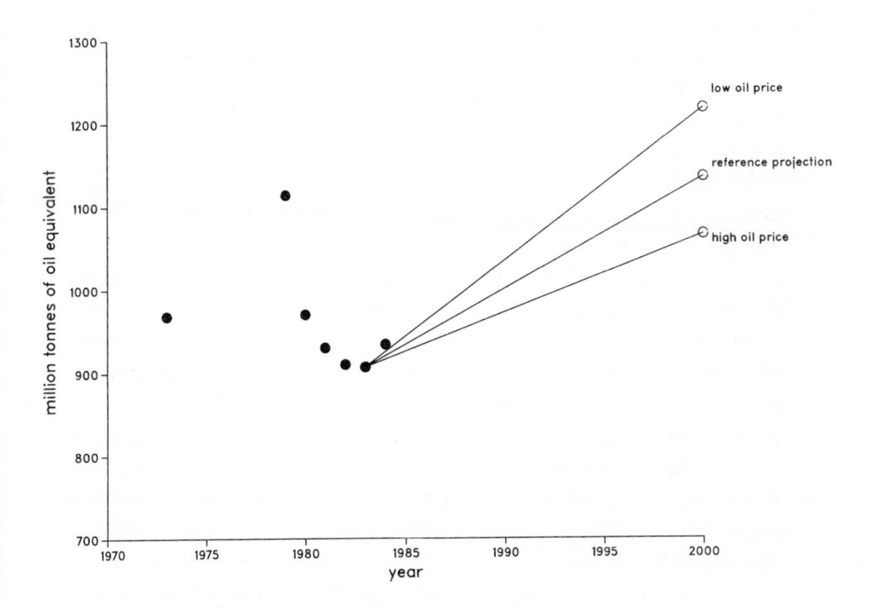

Figure 6.2 Demand projections in *Energy 2000*
Source: Energy 2000

This narrow range of possibilities is further emphasized when the alternative projections for gross energy consumption are examined, in figure 6.2.

Figure 6.2 plots the projections for gross energy consumption under the reference projection, the low oil price scenario, and the high oil price scenario, and in 2000 the range of these projections is from 1066 mtoe (high oil price) to 1218 mtoe (low oil price). This range of 152 mtoe in the projections for the end of the century is, in fact, less than three-quarters of the range of consumption figures in the four years preceding the report's publication. This suggests that only a restricted set of elasticity responses to economic factors has been considered in drawing up the report. Once again, the recent volatility of energy markets is in marked contrast to the narrow range of outcomes considered by the Commission's work.

Turning to the supply side, the clear message of *Energy 2000* is that the Community will replace its pre-1973 oil dependence by a significant switch to nuclear power with a corresponding penetration of electricity into final consumption markets. Such nuclear investments are heavily capital intensive and sensitive to variations in the price of fossil fuel alternatives, especially when the running cost savings of nuclear power are discounted back to present value terms. There may be a strong temptation therefore for member states to retrace the history of the 1960s when indigenous coal resources were protected against falling oil prices.

In the event that the weakness in oil markets caused by the break up of the OPEC cartel is long lived, member states will be faced with the dilemma of protecting nuclear power investments or once again raising consumption of cheaper oil. Certainly, a security of supply premium will already be attached to oil prices in the minds of consumers and producers simply on the basis of the adjustment costs imposed in the 1970s. In 1986, as oil prices fell, there were concerted suggestions in several member states for oil import taxes to discourage a resurgence of oil demand. This dilemma of whether to protect nuclear investments against falling (though possibly only temporarily) fossil fuel prices is only one of the trade offs that might need to be considered.

The Commission has made the reservation that *Energy 2000* 'may be erring on the side of conservatism about the future development of the market for natural gas'. Despite the large potential for increased consumption following the upsurge of the 1970s, and the excess supply generated price renegotiations of the early 1980s, *Energy 2000* is in fact very pessimistic about natural gas. Although production and import volumes are projected to rise, the share of natural gas in gross energy consumption is projected to fall from 19 per cent in 1983 to 18 per cent in 1990 and 17 per cent in 2000.

Behind these projections lies the assumption that while the comparison of nuclear and fossil fuel costs renders nuclear electricity increasingly economic, the security of supply premium implicitly added to long-term contract gas prices from the USSR renders increased gas consumption uneconomic. This presumably also applies to the premium that would be embodied in Community funded development of the Norwegian gas fields towards the Arctic Circle.

Energy 2000 does not explicitly discuss these two comparative economic trade offs: nuclear costs *vis-à-vis* coal or oil compared with the difference between prevailing European gas prices and their security of supply premium. However, given the fluctuations in oil and coal prices in the 1980s, the historic experience of cost overruns and delays in the previous ambitious nuclear programmes, and the availability of a non-

OECD gas supply, this comparison may become more important as time passes.

Energy in Europe: a summing up

If this study had restricted itself to considering politically agreed progress towards a Common Energy Policy for the European Community it would not have filled a single page. Instead, it has examined the development of fuel markets in the EEC over the 1970–86 period, and considered the policy issues that have arisen and been debated both inside and outside the Commission. It is clear that an enormous number of critical and controversial issues are opened up with this broader approach, and the previous chapters have only been able to scratch the surface of the enormous topic of energy in Europe.

The initial discussion highlighted five broad issues in an analytical framework.

The first of these was the macroeconomic response to oil shocks. The 'energy crisis' impinged on the European consciousness principally because it took the form of a monopoly inspired reduction in European living standards. The avoidance of inflation, the signalling of the necessary wage and price adjustments and the support of energy conservation all became concentrated on government use of monetary and fiscal policy. It was clear that responses differed over the 1970s with the initial attempts to cushion the oil shocks abandoned by the late 1970s, with consequent economic recession.

The second issue concerned relations with OPEC and the history of oil in Europe. This has encompassed two trends – the growth of national oil companies but the protection of indigenous coal in the cheap oil era, and, secondly, the strong switch away from oil in the 1970s. The latter began with the dilemma of whether to pursue conciliation or confrontation, but coalesced into straightforward attempts to reduce oil dependence through conservation.

Conservation therefore is the third principal issue in the analysis, and as energy price signals eventually began to work through use of the Community's pricing for rational use of energy, a very substantial change in the energy and oil intensities of real GDP could be observed. This conservation through price signals has evolved into the most effective achievement of energy policy in the EEC, and has worked almost entirely through the market responses of consumers and producers in individual member states.

A fourth issue is the supply response because higher world oil prices not only led to energy conservation but also commercialized much of the Community's indigenous energy production. Nuclear power emerged as the Commission's favoured supply source, as it did in France and to a certain extent West Germany. The economics of nuclear power raised a considerable number of contentious issues, as did the public acceptance of nuclear policies, and these were not fully resolved by the mid-1980s. Natural gas has been a very rapidly growing fuel for the Community but its further expansion has raised questions about security of supply. The Commission has implicitly viewed these as more problematic than the issues of nuclear costs, and favoured nuclear generation of electricity as the replacement for oil dependence.

The last issue is the relation between policy making institutions and the member states. Community energy policy remains simply a collection of individual policies, but it is clear that there is some degree of agreement here on the success of marginal cost based pricing policies in producing considerable conservation.

It is unlikely that the Community will, in the foreseeable future, go beyond the general agreement about conservation directed use of rational pricing in its energy policy making, but it is equally clear that a very large number of unresolved issues remain in European fuel markets. The years following 1986 are likely to be no less interesting than the years preceding it, in which energy in Europe has been dominated by the rise and fall of OPEC.

Notes

Chapter 1 Energy in Europe

1 The quotation is from Mr Mosar's preface to the EC Commission (1985a) publication *Energy in Europe*, April, p. 7.

2 The gross consumption of primary electricity is 91.6 mtoe (table 1.1, row 5) while the final consumption of all electricity is the only slightly larger figure of 98.3 mtoe (table 1.1, row 9) but this does not mean that 93 per cent of electricity comes from primary sources. The final consumption of electricity was 1229 billion kWh which converts to 98.3 mtoe because 1 mtoe has the same energy content as 12.5 billion kWh. However, the primary electricity reported in row 5 is calculated by Eurostat as the amount of oil that would have to be burnt in fossil fuel power stations operating at about 20 per cent efficiency in order to produce the kWh actually generated by nuclear and hydropower. This conversion factor is therefore about five times larger than the strict energy equivalent used in row 9. An excellent guide to the conventions used in energy balance tables is Roberts and Hawkins (1977).

3 The index of export unit values for industrial countries (defined to be OECD excluding Turkey, Greece and Portugal) reported in the International Monetary Fund's *International Financial Statistics* (1981, 1982, 1984a, b).

Chapter 2 The oil shocks

1 This view, that the nature of institutions shapes policy making and the perception of energy problems is a cavalier summary of the approach of Lucas (1977, 1985), a pioneer of energy in Europe studies.

2 These estimates appear in OECD (1980a, 1983a) and Llewellyn (1983).

3 The trade figures in the last two paragraphs are taken from the UK Treasury (1985).
4 The analysis of oil shocks given here draws largely on Corden (1981), Fried and Schultze (1975), Llewellyn (1983) and Powell and Horton (1985).
5 Tables 2.1 and 2.2 summarize the analyses presented in various issues of the OECD *Economic Outlook*, and this is also the source for the definition of structural and inflation adjusted structural budget balances.
6 The current account analysis which follows is based on Sachs (1981).
7 The analysis here draws on Williamson (1983) and the commentary in International Monetary Fund (1984b).
8 See the analysis in Williamson (1983).
9 See 'Oil is cheap today, dearer than yesterday', *The Economist*, 23 November 1985.
10 See the summary in A. Kaletsky 'A free lunch at OPEC's expense', *Financial Times*, 7 February 1986.

Chapter 3 Europe and the world oil market

1 Among the useful oil market analyses used in this chapter are Adelman (1972), EC Commission (1976), Goodwin (1981), Griffin and Teece (1982), Odell (1979), Stocking (1971) and the papers by Adelman, Penrose and Robinson in Hawdon (1984).
2 The analysis which follows draws on Caves and Jones (1981) and the paper by Pindyck in Griffin and Teece (1982).
3 See for example the papers in Griffin and Teece (1982).
4 Discounting for time is an economic appraisal technique which is absolutely fundamental to energy economics. Even in the absence of inflation, possessing current income offers the opportunity of profitably investing it in order to obtain a rate of return denied to those who have to wait to receive their incomes, therefore the rational recipient of an income stream undervalues income that he has to wait for relative to income received immediately. The 'discount factor' is obtained by dividing next year's income flow by 1 plus the rate of return on foregone investment opportunities (r). As income flows recede into the future they are discounted by dividing by $(1 + r)^n$ where n is the number of years which must elapse before receipt. The discounted present value of an income stream is the

sum today of these discounted future incomes. A good reference on this fundamentally important technique is Bierman and Smidt (1971).

5 See Yager's contribution in Goodwin (1981).

Chapter 4 Energy conservation and demand in the European Community

1 See Official Journal C/153 of 9 July 1975, Official Journal C/149 of 18 June 1980 and EC Commission (1985b).
2 Recall the economist's use of the term elasticity: output or income elasticity of demand is the percentage change in consumption resulting from a 1 per cent change in output or income; price elasticity of demand is the percentage change in consumption resulting from a 1 per cent change in price.
3 See EC Commission (1984a). The following survey also draws on the reports by the Economist Intelligence Unit (1984, 1985) and Lucas (1985).
4 The description here draws on EC Commission (1984b), especially the statistical annex, and on the Eurostat (1978, 1984a, b, 1985a, b) (Statistical Office of the European Communities) reports.

Chapter 5 Investment in European fuel supply: policy issues

1 The problems of discounting in energy supply (particularly nuclear and depletable resources) are analysed in Lind (1982) and Jones (1984). See also note 4 of chapter 3.
2 See 'Nuclear safety: pricing a life', *The Economist*, 22 March 1980.
3 See EC Commission (1984c).
4 In particular see Solow (1974).
5 On oil and gas taxes see Devereux and Morris (1983). This whole section draws on OECD (1980b, 1982, 1983b).
6 Good references on which this section draws are the International Energy Agency (IEA) (1982) and EC Commission (1985a).
7 Sources for the data and some of the analysis in this section are the International Energy Agency (IEA) (1982), and the Economist Intelligence Unit (1984, 1985).
8 George (1985) elaborates on this.
9 See EC Commission (1985c).

10 On this topic see EC Commission (1985b).
11 See EC Commission (1985b); International Energy Agency (IEA) (1982) gives useful background on this area of energy supply.
12 The estimates given here are from 'Alternatives to oil', *The Economist*, 26 December 1981.
13 See the Economist Intelligence Unit (1985).

Chapter 6 The evolving energy balances of the EEC

1 See EC Commission (1985a, 1985d).
2 The certainty equivalent oil price is the one which if it were to prevail continuously would not cause the allocation of resources to differ from the risk avoiding outcome observed in uncertain situations, i.e. it incorporates any risk premium implicit in consumers' behaviour.
3 See EC Commission (1985b).
4 See EC Commission (1984c).

Appendix

Indigenous fuel resources in the Community

This Appendix summarizes the characteristics of the distribution of fuel supply resources and consumption in the Community of EUR10 in the first half of the 1980s.

On a country-by-country basis figure A.1 illustrates the relative shares of each fuel type in gross primary energy consumption, while figure A.2 shows consumption of gross primary energy relative to imports for each member state. It is apparent from figure A.2 that the overall import dependence figure of 42 per cent quoted in chapter 1 is heavily influenced by the self-sufficiency of just two member states: the Netherlands and the UK. The subsequent commentary analyses the role of different fuel types in production and imports in each member state.

The production shares of member states are partly the result of the distribution of fuel resources and figure A.3 illustrates the geographical dispersion throughout the Community of the main fossil fuel resources: coal and lignite, crude oil, and natural gas.

Finally, tables A.1 and A.2 show which member states are the leading producers by fuel type and, in particular, notes the role of the net exporters.

In using the figures and tables it is important to note that the primary electricity component, including nuclear, of gross energy production is measured on the oil equivalent generation basis used (following Eurostat) in row 5 of table 1.1, and explained in more detail in note 2 of chapter 1.

Belgium was a traditional coal producer which became dependent on oil and gas imports. It is now switching to nuclear power and 60 per cent of its indigenous energy production is in the form of nuclear-generated electricity.

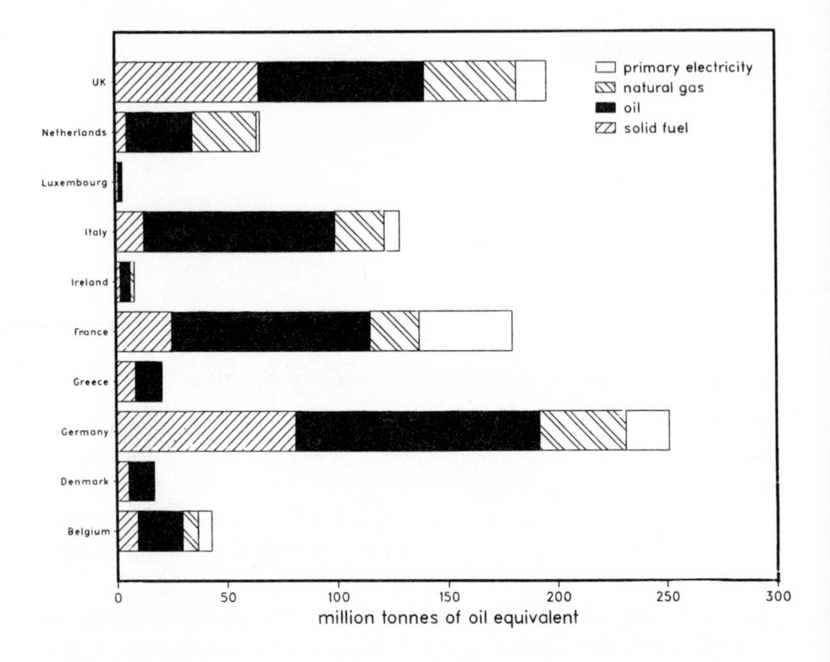

Figure A.1 Primary energy consumption by fuel type, EEC EUR10, 1983

Source: Eurostat

Denmark is also very heavily dependent on imports, with its own North Sea oil accounting for all of its indigenously produced share (13 per cent) of gross primary energy consumption. Its imports are split 40:60 between coal and oil.

Germany has always been one of the Community's leading energy producers, but is now more than 50 per cent dependent on imports (80 per cent oil and 20 per cent natural gas). Its indigenous production is still largely dominated by its hard coal and lignite, but it is also the Community's second largest nuclear power producer. It is the Community's largest solid fuel producer and its second largest total energy producer.

Greece, despite some development of indigenous resources, is still largely dependent on oil imports as it approaches full industrialization.

France has been the most adaptable of the old energy producers switching from indigenous coal to imported oil and then to nuclear power

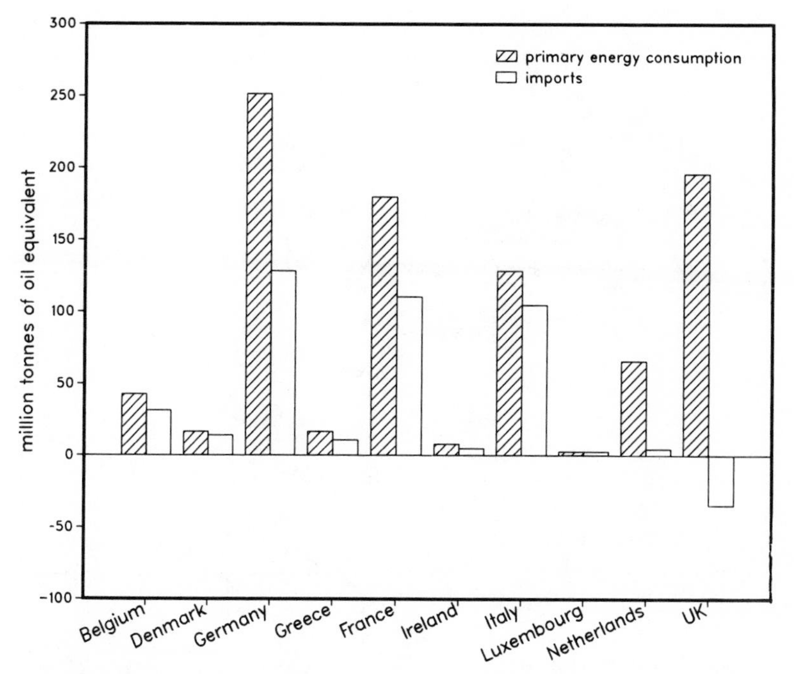

Figure A.2 Primary energy consumption and imports in EEC EUR10, 1983

Source: Eurostat

over a 25-year period. Its huge expansion of nuclear power makes it the Community's leading user of primary electricity, and makes it one of only three member states that are net fuel exporters. This growth in electricity trade through international interconnection could be an important characteristic of future EEC fuel consumption. Nevertheless oil remains dominant in both French imports and consumption, so that with Germany it can greatly benefit from falling oil prices.

Ireland is often categorized like Greece as not yet having reached full industrialization status. Nevertheless it is self-sufficient in natural gas which has overtaken peat as its largest indigenous fuel product.

Italy is still very heavily import dependent, chiefly in oil but also in coal and natural gas. It has yet to exploit fully its planned nuclear potential.

Luxembourg is a minor, import-dependent consumer.

The Netherlands has a very substantial net export surplus in natural

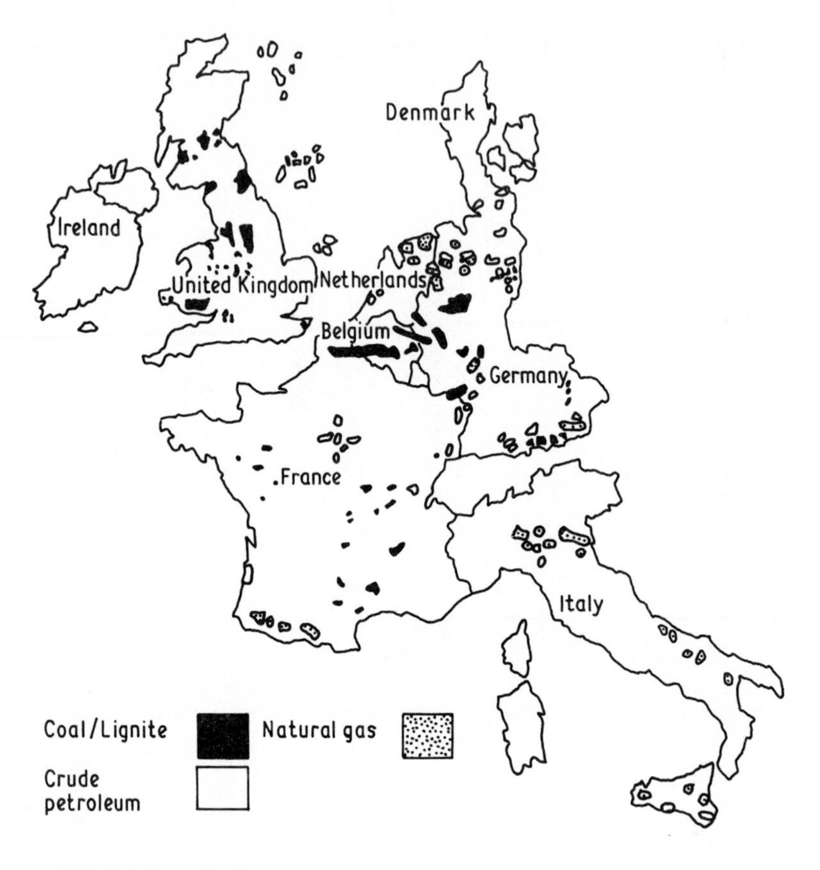

Figure A.3 Indigenous energy resources of the EEC

gas which is just offset by its oil and coal imports so that the country is 90 per cent self-sufficient. Natural gas consumption rivals oil consumption and is used extensively in electricity generation.

The UK is the Community's only net exporter of total primary energy, a role entirely dependent on its having the lion's share of Community oil production. This now exceeds its traditional energy base, coal production, by about 40 per cent. It is the Community's second largest gas producer and its third largest nuclear power user, but its performance in this last regard has been disappointingly short of its initial potential. With France it is one of the Community's leading nuclear fuel reprocessing states.

Table A.1 Leading fuel producers in the EEC by percentage share of indigenous energy production in 1983

Fuel	*Member State (% share)*				
solid fuel	Germany (48)	UK (40)	France (6)	Belgium (5)	others (1)
oil	UK* (88)				others (12)
natural gas	Netherlands* (46)	UK (28)	Germany (12)	Italy (9)	others (5)
primary electricity	France* (42)	Germany (18)	UK (16)	Belgium (7)	others (17)
total gross primary energy production	UK* (45)	Germany (23)	France (12)	Netherlands (12)	others (8)

* Net exporter.
Source: Eurostat.

Table A.2 Gross primary energy production by member state of the EEC, 1983 (mtoe)

Belgium	10.8
Denmark	2.2
Germany	120.6
Greece	5.4
France	63.4
Ireland	2.9
Italy	19.1
Luxembourg	0.1
Netherlands	59.4
UK	232.9
	516.8

Source: Eurostat.

References

Adelman, M. A. (1962) 'The supply and price of natural gas', *Journal of Industrial Economics Supplement*.

Adelman, M. A. (1972) *The World Petroleum Market*, Baltimore, The Johns Hopkins University Press.

Adelman, M. A. (1980) 'The clumsy cartel', *The Energy Journal*, 1, 43–53.

Adelman, M. A. (1984) 'International oil agreements', *The Energy Journal*, 5, 1–9.

Audland, C. J. (1984) 'Energy priorities and options for the European Community', *Atom*, no. 332, 6–8.

Bierman, H. and Smidt, S. (1971) *The Capital Budgeting Decision*, 3rd edition, New York, Collier Macmillan.

Black, R. A. (1977) 'Plus ça change, plus c'est la même chose: nine governments in search of a common energy policy' in H. Wallace, W. Wallace and C. Webb (eds) *Policy-Making in the European Community*, Chichester, John Wiley.

Brondel, G. and Morton, N. (1977) 'The European Community – an energy perspective', *Annual Review of Energy*, 2, 343–64.

Caves, R. E. and Jones, R. W. (1981) *World Trade and Payments*, 3rd edition, Boston, Little, Brown & Co.

Corden, W.M. (1981) *Inflation, Exchange Rates and the World Economy*, 2nd edition, Oxford, Clarendon Press.

Devereux, M. and Morris, C. (1983) *North Sea Oil Taxation*, IFS Report Series No. 6, London, Institute of Fiscal Studies.

Dramais, A. and Thys-Clement, F. (1979) 'The European energy consumption model' in A. Strub (ed.) *Energy Models for the European Community*, London, IPC Science and Technology Press for the Commission of the European Communities.

EC Commission (1976) *Report by the Commission on the Behaviour of the Oil Companies in the Community During the Period from October 1973 to*

March 1974. Studies: Competition-Approximation of Legislation Series no. 26.

EC Commission (1978) *European Documentation: The European Community and the Energy Problem*, 2nd edition.

EC Commission (1983a) *European Documentation: The European Community and the Energy Problem*, 3rd edition.

EC Commission (1983b) 'Energy and the economy: a study of the main relationships in the countries of the European Community', *European Economy*, no. 16, 29–88.

EC Commission (1984a) *Review of Member States' Energy Policies*, COM (84) 88.

EC Commission (1984b) *The Application of the Community's Energy Pricing Principles in the Member States*, COM (84) 490.

EC Commission (1984c) *Nuclear Industries in the Community. Illustrative Programme under Article 40 of the Euratom Treaty (Programme Indicatif Nucléaire pour la Communauté, PINC)*, COM (84) 653.

EC Commission (1985a) Directorate General for Energy, *Energy in Europe*, no. 1, April.

EC Commission (1985b) Directorate General for Energy, *Energy in Europe*, no. 2, August.

EC Commission (1985c) *Energy in Europe*, no. 3, December.

EC Commission (1985d) *Energy 2000. A Reference Projection and its Variants for the European Community and the World to the Year 2000*, SEC (85), 324.

The Economist Intelligence Unit (1984) *Quarterly Energy Review, Western Europe: Annual Supplement 1984*, London, The Economist Intelligence Unit.

The Economist Intelligence Unit (1985) *Quarterly Energy Review, Western Europe: Annual Supplement 1985*, London, The Economist Intelligence Unit.

Eurostat (1978) *Electricity Prices 1973-78*, Statistical Office of the European Communities.

Eurostat (1984a) *Electricity Prices 1978-84*, Statistical Office of the European Communities.

Eurostat (1984b) *Gas Prices 1978-84*, Statistical Office of the European Communities.

Eurostat (1985a) *Electricity Prices 1980-85*, Statistical Office of the European Communities.

Eurostat (1985b) *Gas Prices 1980-85*, Statistical Office of the European Communities.

Fried, E. R. and Schultze, C. L. (eds) (1975) *Higher Oil Prices and the World Economy*, Washington, The Brookings Institution.

George, S. A. (1985) *Politics and Policy in the European Community*, Oxford, Oxford University Press.

Gittus, J. (1986) 'Risk assessment for the pwr', *Atom*, no. 352, 8–10.

Goodwin, C. D. (ed.) (1981) *Energy Policy in Perspective: Today's Problems, Yesterday's Solutions*, Washington, The Brookings Institution.

Grathwohl, M. (1982) *World Energy Supply: Resources, Technologies, Perspectives*, Berlin, Walter de Gruyter & Co.

Griffin, J. M. and Teece, D. J. (eds) (1982) *OPEC Behaviour and World Oil Prices*, London, George Allen & Unwin.

Hawdon, D. (1984) *The Energy Crisis: Ten Years After*, London, Croom Helm.

International Energy Agency (IEA) (1982) *World Energy Outlook*, Paris, OECD/IEA.

International Monetary Fund (IMF) (1981) *International Financial Statistics: Supplement Series No. 2. Prices*, Washington, IMF.

International Monetary Fund (IMF) (1982) *International Financial Statistics: Supplement Series No. 4: World Trade*, Washington, IMF.

International Monetary Fund (IMF) (1984a) *International Financial Statistics: Supplement Series No. 7: Output*, Washington, IMF.

International Monetary Fund (IMF) (1984b) *International Financial Statistics: Supplement Series No. 8: Balance of Payments*, Washington, IMF.

Jones, P. M. S. (1984) 'Discounting and nuclear power', *Atom*, no. 338, 8–11.

Kaser, M. (1985) 'Soviet gas supplies', paper presented at the University of Surrey Energy Economics Centre Conference on International Gas Prospects and Trends, 16 April.

Lind, R.C. (ed.) (1982) *Discounting for Time and Risk in Energy Policy*, Washington, Resources for the Future.

Llewellyn, J. (1983) 'Resource prices and macroeconomic policies. Lessons from two oil price shocks', *OECD Economics and Statistics Department Working Papers No. 5*, April.

Lucas, N. (1977) *Energy and the European Communities*, London, Europa Publications.

Lucas, N. (1985) *Western European Energy Policies: A Comparative Study*, Oxford, Clarendon Press.

Maull, H. (1980) *Europe and World Energy*, London, Butterworths

in association with the Sussex European Research Centre.
Maull H. (1981) *Natural Gas and Economic Security*, The Atlantic Papers No. 43, Paris, The Atlantic Institute for International Affairs.
Nordhaus, W. (1977) 'The demand for energy: an international perspective', in W. Nordhaus (ed.) *International Studies of the Demand for Energy*, Amsterdam, North Holland.
Odell, P. R. (1979) *Oil and World Power*, 5th edition, London, Penguin Books.
Odell, P. R. (1981) 'The energy economy of western Europe: a return to the use of indigenous resources', *Geography (Journal of the Geographical Association)*, 66, part 1, 1–14.
OECD (1980a), 'The impact of oil on the world economy', *OECD Economic Outlook*, 27, 114–130.
OECD (1980b) *Economic Survey: Norway*, Paris, OECD.
OECD (1982) *Economic Survey: Norway*, Paris, OECD.
OECD (1983a) 'Effects of changes in energy prices', *OECD Economic Outlook*, 33, 75–80.
OECD (1983b) *Economic Survey: The Netherlands*, Paris, OECD.
Pearce, D. W. (1979) 'Social cost benefit analysis and nuclear futures', *Energy Economics*, 1, April.
Powell, S. and Horton, G. (1985) 'The economic effects of lower oil prices', *Government Economic Service Working Paper No. 76* (Treasury Working Paper No. 34), London, HM Treasury.
Ray, G. F. (1982) 'Europe's farewell to full employment?' in D. Yergin and M. Hillenbrand (eds) *Global Insecurity* (an Atlantic Institute for International Affairs Study), London, Penguin Books.
Roberts, W. N. T. and Hawkins, W. A. (1977) 'Energy balances – some problems and recent developments', *Energy Paper No. 19*, London, UK Department of Energy, HMSO.
Sachs, J. D. (1981) 'The current account and macroeconomic adjustment in the 1970s', *Brookings Papers on Economic Activity*, 1, 201–281.
Solow, R. (1974) 'The economics of resources and the resources of economics', *American Economic Review*, 64, 1–14.
Stern, J. P. (1984) *International Gas Trade in Europe*, London, Heinemann Educational Books.
Stocking, G. W. (1971) *Middle East Oil*, London, Allan Lane.
Tolley, G. and Wilman, J. (1977) 'The foreign dependence question', *Journal of Political Economy*, 85, 323–47.
UK Monopolies and Mergers Commission (1981) *The Central*

Electricity Generating Board, London, HMSO.
UK Treasury (1985) 'Trends in manufacturing trade and output', *Economic Progress Report*, no. 178, July.
US President's Commission (1979) *Report of the President's Commission on the Accident at Three Mile Island (Chairman John G. Kemeny)*, New York, Pergamon Press.
Williamson, J. (1983) *The Open Economy and the World Economy*, New York, Basic Books.

Index